The Science and Politics of Covid-19

Michel Claessens

The Science and Politics of Covid-19

How Scientists Should Tackle Global Crises

 Springer

Michel Claessens
European Commission
Bruxelles, Belgium

ISBN 978-3-030-77863-7   ISBN 978-3-030-77864-4   (eBook)
https://doi.org/10.1007/978-3-030-77864-4

This Springer imprint is published by the registered company Springer Nature Switzerland AG
The registered company address is: Gewerbestrasse 11, 6330 Cham, Switzerland

Preface and Acknowledgments

The appearance of a single author's name on the cover page is misleading: this book owes a lot to many people—scientists, experts, government representatives, journalists and citizens like you and me—who have contributed to this book, in particular through hundreds of exchanges and interviews, carried out of course by email or videoconference in these Covid-19 times. I want to express my sincere and deep gratitude to all of them.

A very special note goes to my wife Xuling and my son Jon, who have been confronted by a double lockdown during the last few months as I spent most of my free time, days and nights, following and analysing the spreading pandemic. I am also grateful to Kamran Abbasi, Florence Ader, Martin Bauer, Jean-Philippe Béja, Charlotte Belaïch, Leïla Belkhir, Eran Bendavid, Gilles Boëtsch, Philippe Busquin, Bruno Canard, Zachary Cooper, Yves Coppieters, Pascal Crépey, Amy Dahan, Jean-Stéphane Dhersin, Cécile Ducourtieux, Michel Dubois, Jean Durieux, Marius Gilbert, Herman Goossens, Henri Gueguen, Deepti Gurdasani, Emma Hodcroft, Philippe Juvin, Raffi Katchadourian, Shengjie Lai, Jonathan Leake, Jean-Marc Lévy-Leblond, Laurent Mucchielli, François Ouellette, Stefano Paglia, Dominique Pécaud, Guy-André Pelouze, Andrew Pollard, Jonathan Roux, Monique Septon, Alessandro Sette, Sir Adrian Smith, Marcello Tirani, Jean-François Toussaint, Michele Vespe, Patrick Vittet-Philippe, Marc Wathelet, Jean-Pascal van Ypersele, Jean Ralph Zahar, Wenhong Zhang and Nanshan Zhong. Great thanks also go to Elizabeth Trump, who turned the book to real English!

Many thanks also go to Springer and my editor Anthony Doyle for accepting to publish a book analysing "science in politics." Popularising a subject like the Covid-19 pandemic is a great challenge as the information is rapidly evolving and is sometimes just wrong! Even the World Health Organization (WHO) has difficulty coping with the abundant fake news and the proliferating mis- and disinformation. I have therefore done my best to retrieve and doublecheck the right information and present it as objectively as possible, whilst also commenting on it wherever appropriate.

A last remark: the opinions expressed in this book are personal and do not in any way bind the European Commission where I have been promoting the dialogue between science and society or, more modestly, between scientists and citizens for many years.

Should you wish to be kept informed about the evolution of this subject, kindly follow me on Twitter @M_Claessens or email me at michel_claessens@yahoo.fr.

Bruxelles, Belgium   Michel Claessens

Introduction

Time cures the sick man, not the ointment

Traditional Proverb

It came out of the blue. On January 2, 2020, WHO issued a global alert on the emergence of several cases of "pneumonia of unknown cause" in Wuhan, China. Except in Beijing, most governments do not show any panic and merely set up daily monitoring of the incubating epidemic. Donald Trump mocks the "Chinese virus" and claims that "everything is under control."

A few weeks later, in March 2020, the coronavirus epidemic becomes a global pandemic, taking almost everyone by surprise. The whole planet experiences a health tsunami and a media pandemic, slowing down economic, diplomatic, educational, sport and cultural activities as well as the daily lives of billions of people for several months. With public health measures imposed in almost every country in the world, including border closures, mandatory teleworking and physical distancing, most capitals are now ghost towns. Masked mankind is mobilising against a coronavirus declared public enemy number one and embarks on a silent world war against an invisible aggressor. Well, a small war actually as Covid-19 has caused today[1] around ten million deaths—0.15% of the world's population—ten times as much as all road accidents combined: a dramatic toll, of course, but numerically much more

---

[1] Unless explicitly mentioned, the data provided in this book refers to June 21, 2021, when I gave the final green light to the printer.

modest than other recent epidemics. The impact is nevertheless terrible from an economic, social and human point of view. In the United States, one in three Americans has lost someone to the coronavirus.

This story is about a collective failure. As Rudolf Virchow, the great nineteenth century father of pathological anatomy, liked to say: "An epidemic is a social phenomenon that has some medical aspects." Hesitating between collective immunity and individual restrictions, most governments turned out to be helpless and indecisive. What was so special about the twenty-first century's worst pandemic—so far—that we lost our confidence and gave in to fear? And who is responsible for this? Governments, who failed to appreciate the risk which was growing from the end of 2019? Scientists, who failed to sound the alarm and bring us to our senses? Doctors, some of whom are today mere merchants of traditional remedies[2]? Or all of us given that we were still mocking the small "flu" in those early months? Is it our technoscientific culture that drives us to always adhere to the myth of *zero risk*?

The fact is that our institutions have failed tremendously. And this collective failure is not just limited to the fight against the virus. The lockdown ideology imposed social distancing with all its painful ramifications, and for many of us the crashing economy is having the same effect. The scientific evidence shows that Covid-19 is stressful for people, increases fear and anxiety, and causes a deterioration in mental health.

How did we get there? Most governments have appointed top-level scientists and world-renowned doctors to advise them. Lots of expert committees have been set up to strengthen the "evidence-based" character of decisions. Indeed, political leaders have repeatedly said that their decisions are based on science. This seemed to be the best approach, given the novelty of the virus, the complexity of the situation and the need to take into account the latest virological and epidemiological knowledge. Thus, science and media stars performed almost every evening on TV shows and on social media to explain the progression of the virus and also, by the way, the progress of our ignorance. The forecasts circulating in government circles and the lessons learnt from China, which was acting as a guinea-pig, could have been useful. However, supported by billions of dollars, the research community quickly embarked on the quest for the Holy Grail—a vaccine against SARS-CoV-2. It is true that no one would have expected a technological society like ours to give up in the face of a "natural" and microscopic enemy. However, the success did not come. What happened? What went wrong?

---

[2]In this book, I consider medical doctors and physicians as scientists, although in most cases they represent different professions and jobs.

Meeting the challenge seemed to be child's play, given the level of our technoscientific knowledge and the power of the pharmaceutical industry. We were wrong: a new virus can turn off the entire world economy. Supported by scientific progress, we have convinced ourselves that science can push back all frontiers and all diseases, and that rich countries have definitely rid themselves of infectious and contagious diseases. This was the first mistake. And a clear message to the research community: scientific information must target both the achievements of knowledge and the limits of science. The fact is that coronaviruses are not at the top of scientists' research priorities.

Thus, during the first months of 2020, science plunged into politics. And scientists found themselves at the heart of—as well as an actor in—a health crisis that brought into the spotlight these improbable pairs of scientists and politicians. Anthony Fauci and Donald Trump in the United States, Chris Whitty and Boris Johnson in the United Kingdom, Jean-François Delfraissy and Emmanuel Macron in France, Nanshan Zhong and Xi Jinping in China: how did these impossible couples manage and survive the crisis? How did scientists and decision-makers live under the same roof even though everything else separates them? Between the long term of science and the too short term of politics, between scientific rigor and political judgment, between the questions of researchers and the responses of decision-makers is there an opportunity for cooperation—and even for mutual understanding? The question is not new and gets asked with each major crisis. Researchers struggle to get the facts and their part of the truth heard, while politicians do not always succeed in sharing their reasoning. Of course, this unique pandemic is now a textbook case, both by the number of countries concerned and by the scale of the measures taken as more than half of humanity was on total lockdown in March 2020. So, what did we learn?

I worked with dozens of scientists, doctors and government officials on the following question: how can the scientific community better interact with the political world to guide decisions without losing its integrity and giving up its *raison d'être*? My contacts, in particular in France, China, the United Kingdom and the United States, allowed me to better understand how these countries have handled the crisis—and hence managed the related uncertainties and fears. One of the most striking characteristics of this pandemic is how little knowledge and certainty we had—and still have. So how can the science community adequately advise decision-makers without being in possession of the full facts themselves? Ironically, governments which claimed to seek and rely on scientific advice, such as France, the United Kingdom and the United States, did not perform any better than other countries—on the contrary even. After so much time spent in crisis, have experts and policymakers succeeded in rationalising politics and politicising science in the best possible ways? Science in *the* crisis or science *in* crisis: what lessons can be drawn from this pandemic, a unique period of history which showcases science policy on a global scale?

Unfortunately, if major successes have been achieved, science apparently failed. There is no specific antiviral remedy yet and even if vaccines are now available, there is still no assurance that a long-lasting immunity will be conferred to the world's population—this will take months or potentially years to answer. Furthermore, many hiccoughs happened in the scientific community: downplaying of the outbreak, wrong vaccine predictions, peer-reviewed publications retracted after a few days, short-circuiting of clinical trials, research supported for political reasons, scientific vagaries, spectacular announcements in the media, etc. We have seen the best as well as the worst in recent months and the public discovered "science by press release." More worryingly, experts themselves acknowledge that they did not rise to the challenge. Those who advised governments failed to have a decisive influence on the timing and the direction of political decisions. Could this be down to a lack of firmness or confidence on their part? Is it because scientists are known to be reserved and place great importance on fact-checking? Or is it because, having seen that politics is what it is, they seemingly resigned themselves to having little impact? We will try to answer these questions. In any case, in the four countries analysed here, the scientific advisers failed to appreciate the risks and the urgency. Apart from China, all these governments wasted time before going into lockdown in early 2020 although this was at that time the only possible option to slow down the pandemic.

Admittedly, in order to respond to the multiple pressures and challenges, the research community has been squeezed and pushed to its limits. The public has been watching, almost *live*, both the advances and the sluggishness of research, together with the progress of both our knowledge and ignorance in the area of epidemiology and infectious diseases. The layperson was introduced to the backstage of research, discovering the behind-the-scenes machinery of science.

In this book, we will discover this "hidden side of science" and see how scientists are working, struggling and even fighting to get professional recognition. Today more than ever, knowledge is power! This unofficial and "unauthorised" presentation of science in the making is a unique opportunity to discover the multiple influences, often complex, sometimes ridiculous, actually very human, which determine the destiny of scientific ideas. It also confirms that, for better or worse, science today is a political business.

Covid-19 has highlighted the *politicisation* of science but we should acknowledge that science is also *politicised*, as shown in particular by the indulgence of the experts with regard to the mistakes made by governments during the pandemic and the silence of the medical community about the risks of treatments and vaccines against Covid-19. Of course, politicians are exploiting science, but scientists are also exploiting politics to promote their own work and ideas. In fact, science has always been inseparable from politics. Covid-19, through a huge mediatisation of scientists and experts, has seen the development of "science politics" at an unprecedented rate.

Before beginning this fascinating story, let me say a few words about this book. To give credit where credit is due, SARS-CoV-2 is the main theme of Chap. 1, which is basically an account of what we do not know about the coronavirus. Next comes a "coronavirus diary," which recalls the main stages of the discovery and the spread of this brand-new infectious agent. Chapter 3 presents the geopolitics of the coronavirus and compares the strategies implemented by the countries which have been on the front line from the start. Finally, we will meet the experts and communicators who are centre stage in this crisis and therefore the best placed to shed light on this hidden side of science. And we will end up with this question: what should we do next? How can we be better prepared for the crises that will follow—which we are certain will come? For sure we are missing an adequate dialogue between scholars and decision-makers and by dialogue, I mean a genuine two-way communication. But there are other factors to consider. Were we wrong to assume from the outset that science could play a decisive role in solving this health crisis? In other words, should scientists assert themselves more ostensibly—and perhaps more honestly? Or, on the contrary, should science be more modest? It would be naive to rely only on scientists to find the solution of the Covid-19 global crisis. After all, managing a lockdown is more a matter of police decisions than a scientific issue. The development of messenger RNA vaccines has been hailed as a scientific achievement. This is partly true. But this breakthrough also reflects the fact that the pharmaceutical industry is researching in areas where sickness rhymes with *healthy* business...

Here is, in my opinion, the real question for us today. What role can scientists play—or not—in these health and environmental crises which we are told will be more frequent and more intense in the future? Some believe that the Covid-19 crisis is a kind of rehearsal of the climate crisis that we will have to face in the future. I don't think this is correct, even if there are some similarities between these issues. But one thing struck me: 195 countries have ratified the Paris Agreement but all have turned in on themselves during the Covid-19 crisis. They all pulled down their blinds, locked their doors and promoted national approaches rather than international cooperation. This has been a major political mistake. The virus ignores borders, just like knowledge. To fight a common enemy, solidarity and cooperation are key. However, scientists have been drawn into sterile competitions and political calculations. Due to a lack of overall collaboration, major projects were cancelled. Fortunately, a huge wave of solidarity and generosity arose in all countries: we all remember medical and nursing staff working days and nights, less rewarded jobs mobilised against all odds to take care of the poor and help the elderly, and volunteers appearing from everywhere to help the disadvantaged and vulnerable. However, this practice did not generally develop at global level, with a few exceptions. The richest countries were keen to pre-order millions of doses of vaccine candidates, failing to show any consideration for the needs of the developing ones. At this point, let me mention a small personal note: I had proposed in early March 2020 to Madame von der Leyen, President of the European Commission, that her services (my colleagues in fact) could stimulate cross-border cooperation in the field of health care and allow the transfer of patients contaminated by the coronavirus to other Member States to ease the pressure on hospitals that were going to be overloaded. The European Commission formalised this initiative on April 3, 2020.[3] As I am a good soldier, I decided to engage in *World War III*...

An opinion survey carried out in France in April 2020, during the full lockdown, revealed a loss of 10 points in the public's confidence in scientists.[4] Researchers are therefore losing some feathers. And one last point about modesty: there is a huge project which is left behind: science education. Let's be honest: we all contributed, more or less, to spreading wrong information and nonsense about epidemics, exponentials, vaccines or lockdowns. Ignorance is prime ground for the growth of the virus. Scholar's dream or technoscientific realism: will we be able to win this fight? Politics and the pharmaceutical industry are the big winners in this crisis.

---

[3]European Commission (2020, April 3) Coronavirus: Commission encourages and facilitates cross-border treatment of patients and deployment of medical staff, https://ec.europa.eu/commission/pressc orner/detail/fr/ip_20_590.

[4]Dubois M (2020, June 18) La crise a-t-elle changé notre regard sur la science, CNRS Le journal, https://lejournal.cnrs.fr/billets/la-crise-a-t-elle-change-notre-regard-sur-la-science.

Contents

1

A Microscopic Killer

There is no cure for an unknown illness.
Myanmar Proverb

Still unknown a few months ago, SARS-CoV-2 (severe acute respiratory syndrome coronavirus 2) is the virus responsible for Covid-19,[1] a mild respiratory disease in 80% of cases, as the current data suggests, but very symptomatic for 15% of contaminated patients and very severe for 5% of them. What do we know about this coronavirus, which apparently appeared for the first time in 2019 in the city of Wuhan, China? Let's take stock of the current knowledge.

A virion, the infectious form of a virus, is a microorganism that parasitises a living cell to "feed" on its energy and in which it will "copy" itself millions of times. SARS-CoV-2 belongs to the coronavirus family, the virions of which have a crown-like appearance under the electron microscope, which is why the viruses are named after the Latin word corona, meaning "crown" or "halo" (see Fig. 1.1). The lastborn of the family is the seventh coronavirus known to infect humans (HCoV). The other six are notably responsible for common colds and digestive infections as well as more serious illnesses such as the 2002–2004 SARS outbreak, an epidemic involving severe acute respiratory syndrome (SARS) which was, as Patrice Debré and Jean-Paul Gonzalez put it,

---

[1]Throughout the book, SARS-CoV-2 refers to the virus and Covid-19 to the related disease.

© The Author(s), under exclusive license to Springer Nature Switzerland AG 2021
M. Claessens, *The Science and Politics of Covid-19*,
https://doi.org/10.1007/978-3-030-77864-4_1

Fig. 1.1 Computer generated image of the SARS-CoV-2 virion. Its capsid is made of several types of proteins and contains one strand of RNA (credit https://www.scient ificanimations.com/wiki-images/, CC BY-SA 4.0)

the "first great fear of the twenty-first century.[2]" Now famous, coronaviruses were until recently unknown to the general public and are studied by only a few scientific teams around the world.

The genetic material of SARS-CoV-2 is contained in a single strand of RNA (ribonucleic acid), which is 29 903 nucleotides long and encodes for just 10 genes,[3] making it the longest known RNA virus genome. The genetic sequence is 79.5% identical to that of the SARS coronavirus. For comparison, the size of the human genome is approximately 3.2 billion pairs of nucleotides. The genetic information contained in the SARS-CoV-2 genome is therefore very limited—about a hundred thousand times smaller than in humans—and does not allow the virions to be autonomous. This is why, in order to survive and develop, the coronavirus needs to contaminate a living cell and exploit its biochemical and biological machinery for its own benefit and allow for the replication of its genome as well as the production of its molecular components (proteins, envelope, etc.).

The new coronavirus was first referred to by WHO as "2019 novel coronavirus," or "2019-nCoV," before receiving its official name, "SARS-CoV-2," on February 11, 2020, following the recommendation of the International Committee of Taxonomy of Viruses (ICTV). The disease caused by the new

---

[2]Debré P and Gonzalez JP (2013) Vie et mort des épidémies, Odile Jacob, Paris.
[3]The nucleic acids (DNA and RNA) are made of four different types of nucleotides. They are present in all life-forms on Earth and contain the information required to direct the biosynthesis of proteins.

virus is called "Covid-19" from the English "coronavirus disease 2019") also since February 11, 2020 (as often, the agent and the disease which it causes have different names).

A "Chinese Virus"?

According to the present data, the new coronavirus made its first appearance at the Huanan Seafood wholesale market in Wuhan, Hubei Province of China, currently designated as the initial source of the outbreak. This suggests that SARS-CoV-2, like the SARS coronavirus, SARS-CoV-1, is of animal origin and has jumped to humans by breaking down one or several interspecies barriers. This was confirmed by the phylogenetic and genomic analysis of the virus, from which it was possible to build its "family tree." There seems to be broad consensus that bats were the host organisms for SARS-CoV-2. However, the Wuhan market did not have bats on offer, as these are not the subject of culinary traditions in Hubei, contrary to what was claimed in some videos that have gone "viral" … According to the latest scientific evidence, SARS-CoV-2 was likely transmitted from a bat to an intermediate host and then spread to humans. A team led by Shi Zhengli, a coronavirus virologist working at the Wuhan Institute of Virology (where she returned in 2000 after having obtained her doctorate from the University of Montpellier in France), showed that the genetic sequence of SARS-CoV-2 is 96.2% similar to a virus usually found in bats and 79.5% similar to that of the SARS coronavirus. Research suggests that the latter is passed to humans via the civet, of which one of the coronaviruses differs from SARS-CoV-1 by only ten nucleotides. By analogy, many scientists now suspect that one or several hosts, still unknown, are the intermediate animals of SARS-CoV-2 between bats and humans.[4] It has also been established that the coronavirus responsible for MERS (Middle East Respiratory Syndrome) also originates from bats and has been transmitted to humans by camels. The pangolin has been considered for a time as a possible host, but the pangolin coronavirus sequences are too different from SARS-CoV-2. In addition, in China, pangolins and bats do not share the same geographic areas. While the Wuhan market may have played a decisive role in the spread of the epidemic, it is most likely not the genuine source. According to some studies, viruses similar to SARS-CoV-2 have also been found in bats living in Burma, Laos and Vietnam. It is not impossible that the virus had already been circulating for several

---

[4]Cohen J (2020, January 31) Mining coronavirus genomes for clues to the outbreak's origins, Science, https://www.sciencemag.org/news/2020/01/mining-coronavirus-genomes-clues-outbreak-s-origins.

years in Southeast Asia. This is the conclusion of the international team of experts from WHO who spent four weeks, from January 14 to February 10, 2021, in Wuhan, to determine the origin of the virus. Early data suggests that Covid-19 could have been circulating for weeks before it was identified in the city of Wuhan in December 2019, according to the experts. Peter Ben Embarek, head of the delegation, concludes that it is "extremely unlikely" that the virus emerged as the result of a lab-related incident. The exact origin of SARS-CoV-2 remains unknown, the final report concludes.[5]

However, eighteen scientists claim in a recent letter published by Science: "Theories of accidental release from a lab and zoonotic spillover both remain viable."[6]. Therefore, SARS-CoV-2 is in no way a "Chinese virus" according to Bruno Canard, a virologist and research director at the French CNRS (Centre National de la Recherche Scientifique), who argues that the ongoing global anthropisation of the world favours the emergence of new viruses that were until now confined in animals and maintained in their natural habitats.[7] Biodiversity seems to be the best defence against interspecies transmission of viruses. This has been confirmed by a recent publication by American researchers, who have shown that endangered wild species (due to human exploitation and loss of habitat) share more viruses with humans.[8] We might have some responsibility in the appearance of new viruses from the animal world.

Many theories about the origin of the SARS-CoV-2 are circulating on social media. Some argue, for example, that the new coronavirus escaped from the very high security laboratory of the Wuhan Institute of Virology (BSL-4, biosafety level 4). According to an article published by *Nature Medicine* in March 2020, several genetic markers in the SARS-CoV-2 genome demonstrate its proximity to other known viruses and its probable evolution from them.[9] There is no need to argue that the virus was made by genetic manipulation: the genome of SARS-CoV-2 does not look like a laboratory construct or a purposefully manipulated virus. The idea is also circulating that "patient zero," or the index case, that is to say the first person

---

[5]WHO (2021, March 30) WHO-convened global study of origins of SARS-CoV-2: China Part, https://www.who.int/health-topics/coronavirus/origins-of-the-virus.

[6]Bloom JD et al. (2021, May 14) Investigate the origins of COVID-19, Science, https://science.sciencemag.org/content/372/6543/694.1.

[7]Culot A (March 13, 2020) La science fondamentale est notre meilleure assurance contre les épidémies, CNRS Le journal, https://lejournal.cnrs.fr/articles/la-science-fondamentale-est-notre-meilleure-assurance-contre-les-epidemies.

[8]Johnson CK (April 8, 2020) Global shifts in mammalian population trends reveal key predictors of virus spillover risk, Proc. R. Soc. B. 287:20,192,736, http://doi.org/10.1098/rspb.2019.2736.

[9]Andersen KG et al. (March 17, 2020) The proximal origin of SARS-CoV-2, Nature Medecine, https://www.nature.com/articles/s41591-020-0820-9.

infected by the virus, appeared in the United States and would have been transferred to China on the occasion of the 7th Military World Games which were held in Wuhan from October 18 to 27, 2019.[10] A senior official of the Chinese Ministry of Foreign Affairs, Lijian Zhao, defends this hypothesis on the basis of a video, reproduced on his Twitter account,[11] of the intervention of Robert Redfield, the then director of the famous U.S. Centers for Disease Control and Prevention (CDC), before a congressional committee on March 11, 2020. In this video, the director acknowledges that people who died of influenza at the start of the epidemic might actually have been victims of Covid-19, without giving a precise time reference. In Europe, several French, Belgian, Luxembourgish and Swedish athletes admitted that they experienced, after their return from Wuhan, severe flu symptoms similar to those described by Covid-19 patients in March and April 2020.[12] However, there is no scientific evidence that they were infected by SARS-CoV-2.

At molecular level, the genome of the novel coronavirus was first decoded on January 5, 2020 by a team from Shanghai's Fudan University, which uploaded the complete sequence on January 11; a technical achievement which has been made possible by advances in genomics and in sequencing machines capable of determining the composition of the genetic material of a living organism. Several strains of the virus have been identified. According to a study published in the prestigious journal *Cell*, SARS-CoV-2 is indeed evolving and it has been proved that a mutation called "D614G" makes it more infectious, but not more dangerous, than the original strain that circulated since late January 2020.[13] Many other mutations have been subsequently identified.

Although microscopic, the SARS-CoV-2 virion is relatively large (0.125 µm in diameter), larger than influenza and SARS virions. The SARS-CoV-2 "antennas" immediately caught the attention of scientists. They are made up of a protein, called "S" (for "spike"), which is a characteristic of SARS-CoV-2. The S protein binds the angiotensin-converting enzyme 2 (ACE2) which is present on the surface of our cells and enables the virus to invade our body. Scientists noted that the virus' binding to the ACE2

[10]Westcott B and Jiang S (March 14, 2020) Chinese diplomat promotes conspiracy theory that U.S. military brought coronavirus to Wuhan, CNN, https://edition.cnn.com/2020/03/13/asia/china-corona virus-us-lijian-zhao-intl-hnk/index.html.

[11]Check-out the message posted on his Twitter account @zlj517 on March 12, 2020.

[12]Picard L and Guyot M (May 6, 2020) Athlète de la Gendarmerie, Elodie Clouvel dit avoir contracté la Covid-19 lors des Jeux militaires de Wuhan, L'Essor de la gendarmerie nationale, https://lessor.org/a-la-une/athlete-de-la-gendarmerie-elodie-clouvel-dit-avoir-contracte-le-covid-19-lors-des-jeux-militaires-de-wuhan/

[13]Korber B et al. (July 2, 2020) Tracking changes in SARS-CoV-2 Spike: evidence that D614G increases infectivity of the Covid-19 virus, Cell, https://doi.org/10.1016/j.cell.2020.06.043.

receptor, through the S protein, is much stronger than for other coronaviruses, which would explain why Covid-19 is spreading much faster and doing a lot more damage in the infected organism. It may also explain why SARS-CoV-2 is also easily transmitted from person to person (for the virus literally "sticks" to our cells). After binding to ACE2, furin, an enzyme ubiquitous in living organisms, transforms the S protein and allows SARS-CoV-2 to enter the host cell to which it is attached. This takes just fifteen minutes on average. There, the virus will feel "at home," safe and protected against external attacks. It will then multiply (typically in several thousand copies) until it causes the death of the host cell. This ubiquitous mechanism explains why, in some infected people, the coronavirus creates damage in very many organs.[14] The genetic sequence encoding protein S is not found in any database. It is unlikely therefore to have been conceived by a laboratory—yet another confirmation of the natural origin of SARS-CoV-2.

How Deadly is the Virus?

In an infected person, SARS-CoV-2 causes in particular an excessive response of the innate immune system as well as respiratory and neurological illnesses. The new coronavirus has the particularity of being able to infect both the upper respiratory tract (nose, nasal cavity, mouth and pharynx) and the lower one, located in the rib cage (larynx, trachea, bronchial tree and lungs) where, in the most severe forms, the virus infects the pulmonary alveoli and causes deadly levels of lung inflammation.[15] This creates a so-called "cytokine storm," a massive production of proteins which results in calling on the macrophages of the immune system. The alveoli, literally flooded with viruses and cytokines, then see their walls thicken, which leads to a significant reduction in oxygen exchange (and hence a severe acute respiratory syndrome—SARS). Some of these patients will feel short of breath and will need breathing and other organ support in Intensive Care Units (ICUs). The virus causes lung lesions, sometimes very quickly, visible on images taken by an X-ray scanner and confirmed at autopsy by pathologists. But the virus also attacks blood vessels and especially their walls. Here too, the virions bind to the ACE2 receptors and then disrupt the cells, the destruction of which

---

[14]Yong E (March 20, 2020) Why the Coronavirus Has Been So Successful, The Atlantic, https://www.theatlantic.com/science/archive/2020/03/biography-new-coronavirus/608338/

[15]Chowdhury MA et al. (July 14, 2020) Immune response in Covid-19: A review, Journal of Infection and Public Health, Vol. 13, Issue 11, 2020, pp. 1619–1629, https://doi.org/10.1016/j.jiph.2020.07.001.

will cause blood clotting. This means that dangerous clots form in the lungs, heart, intestines or brain. The vessels become blocked; the blood thickens: all changes that increase the risk of a fatal outcome.

Unlike other respiratory viral infections, the response to Covid-19 covers a very broad clinical spectrum. The virus can attack the liver, kidneys, heart, vascular system and, in fact, almost every organ in the body: for example, some patients have suffered from paralysis of the tongue which prevented them from speaking for several days. Other patients have experienced persistent hiccoughs, headaches, nausea, loss of smell—anosmia—and/or taste—ageusia. Another important phenomenon: many patients do not regain their usual state of health until after several weeks following the positive test for Covid-19. This is confirmed by a scientific study published in January 2021 on a cohort of 1 700 Chinese hospitalised in spring 2020: 76% of them still suffer, six months later, from at least one symptom.[16] They suffer from real physical and mental exhaustion which has a profound impact on daily life. Some describe their experience as if they have been hit by a truck, were completely drunk, or had severe jet lag—or all three together.

The incubation period for SARS-CoV-2 (from contamination by the virus to the first symptoms) is estimated to be between 2 and 10 days,[17] with a median value of 5 days (which means that half of the patients experience symptoms before 5 days). Contrary to other viral diseases, people infected by SARS-CoV-2 can be contagious before the first symptoms appear.

However, it should be remembered that between 5 and 80% of people tested positive for SARS-CoV-2 may be asymptomatic or presymptomatic: they appear healthy and mostly recover without special treatment, but they nevertheless transmit the virus. In these cases, the immune system may handle the infection wonderfully, without the infected person even knowing it. In some places, for reasons still unknown, the number of asymptomatic cases turns out to be particularly high, for example, in the Italian village of Vò in Veneto, where the first Covid-19 patient died in Italy on February 21, 2020. As nearly all the 3 200 inhabitants of the village had been confined and tested at the beginning and at the end of the lockdown (25 February and 7 March), scientists came to the conclusion that 42.5% of the confirmed infections were

---

[16]Huang C et al. (2021, January 8) 6-month consequences of Covid-19 in patients discharged from hospital: a cohort study, The Lancet, https://www.thelancet.com/journals/lancet/article/PIIS0140-673 6(20)32656-8/fulltext.

[17]However, in July 2020, 57 Argentinian sailors were infected after spending 35 days on a fishing boat and after completing 14 days of mandatory quarantine. They all tested negative before leaving. No scientific publication, however, reports such a long incubation period.

asymptomatic.[18] They also showed that their quarantine reduced the number of infections (prevalence) by 50% in two weeks. As of May 28, 2020, three inhabitants of Vò had succumbed to Covid-19.

What do we know about the risk of dying from Covid-19? The probability can be estimated from the proportion of infected individuals with fatal outcomes. This ratio is called the infection fatality rate (IFR) and requires a complete picture of the number of infections of, and deaths caused by, the disease. The IFR for Covid-19 is estimated to be around 1–2%.[19]

However, in most cases, we don't know the exact value of the number of infections because in general not everyone is tested. This is why two other metrics are used in practice. The first is the Covid-19 case fatality rate (CFR), which is the number of confirmed deaths divided by the number of confirmed cases (i.e., any person meeting the laboratory criteria—detection of virus material, antibodies, etc.). This rate increases with the age of the patient, varies substantially across different locations for complex reasons, and typically changes over time. You can easily calculate the CFR from the available data, and you will get for example 9% for Mexico, 3% for the United Kingdom, 2% for Germany, the United States and France and 1% for New Zealand.[20] These values are higher than the IFR because in all these countries the number of people tested positive underestimates the total number of infections. In most countries, CFR values decrease over time because testing capacities increase and doctors are now better at treating Covid-19 patients than they were in January 2020.

The second metric is the crude mortality rate, sometimes called the crude death rate, which measures the probability that any individual in the population will die from the Covid-19, and is calculated by dividing the number of deaths from the disease by the total population. This rate also varies across countries and is increasing over time. Currently, the highest mortality rates are found in Peru and Hungary (respectively around 0.6% and 0.3% of the population). These theoretical considerations show that it is difficult to compare countries: they may have different strategies to detect and report

---

[18]Lavezzo E et al. (2020, June 30) Suppression of a SARS-CoV-2 outbreak in the Italian municipality of Vò, Nature, https://doi.org/10.1038/s41586-020-2488-1.

[19]You will find in the literature other values for the infection fatality rate of Covid-19. This is due in part to the uncertainty about the number of infected cases. What happened on the *Diamond Princess* cruise ship in February 2020 is interesting because monitoring the evolution of the disease in a closed population was possible. Of the 3,711 passengers and crew, 712 people tested positive (as of March 22, 2020), 128 were asymptomatic and 7 died. The infection fatality rate is therefore 1%.

[20]Ritchie H et al., Mortality Risk of Covid-19, Our World in Data, https://ourworldindata.org/mortality-risk-covid.

Covid-19 deaths and may be using different case definitions, testing strategies and counting methods (for example, mild cases not being tested or counted).

How Can We Protect Ourselves?

As of January 20, 2020, the Chinese pulmonologist Nanshan Zhong, responsible for the discovery of SARS and world-renowned expert in respiratory diseases, confirms human-to-human transmission of SARS-CoV-2. It is therefore essential to know a fundamental characteristic of the infection, namely its *basic* reproduction rate, also called factor R_0 (pronounced R nought or R zero), that is to say the expected average number of people infected by one case in a population where all individuals are susceptible to infection. This factor indeed determines the contagiousness of the virus and the number of people who, on average, will be infected once the epidemic has spread. Virologists and epidemiologists working with the coronavirus have estimated the R_0 factor of SARS-CoV-2 to be around 3, in the absence of any social distancing. This means that on average three people will be infected from one case. However, R_0 is not a biological constant for SARS-CoV-2 as it is affected by other factors such as environmental conditions and the behaviour of the infected population. This leads to the *effective* reproduction number (usually written R_t, with t for time), which is the number of cases generated in a population at time t.

The infection of a given population by SARS-CoV-2 follows a geometric progression (the discrete version of the exponential law): one carrier infects on average three other carriers in 5 days (the estimated "serial interval," i.e., the time from illness onset in a primary case—infector—to illness onset in a secondary case—infected), nine other people in the next 5 days, then 27, 81, and so on. In a month, that represents nearly a thousand people infected. It's a genuine chain reaction. As in an atomic bomb, when a uranium nucleus splits under the impact of a neutron, it in turn releases three neutrons, which in turn will split up three other nuclei and so on. Hence the importance, in the absence of any pharmacological treatment and vaccine for Covid-19, to act at the start of the epidemic and quickly impose "social distancing" or, even better, "physical distancing" in order to reduce contact and exchanges between exposed individuals. On the other hand, researchers have also highlighted the existence of "super-contaminators," patients who, for some as yet unknown reason, are capable of transmitting the coronavirus to dozens or even hundreds of people.

Although geometric progressions are taught at school, few of us realise the inexorability of the exponential evolution. In the case of Covid-19, it means that it can get out of control very rapidly. And even if you impose a total lockdown, you will have to wait a minimum of two or three weeks before perceiving any improvement. As German scientists have demonstrated, we tend to believe that the evolution of the epidemic follows a linear trend.[21]

SARS-CoV-2 is transmitted mainly through respiratory droplets emitted when we speak, cough or sneeze. Using a laser beam to visualise these bursts, American researchers have shown that a speaker can emit up to ten thousand oral fluid droplets per second and that they can remain suspended in the air of a confined place for up to fourteen minutes.[22] Two Cypriot scientists, for their part, have simulated the dispersion of a cloud of droplets emitted by a sneeze.[23] Downwind, these can travel airborne up to six meters! However, the cloud of droplets is reduced to nothing if the emitter wears a mask. These articles provide a scientific demonstration of the mask's utility in limiting the spread of the coronavirus. Other scientific publications indicate that surgical face masks prevent transmission of human coronaviruses and influenza viruses from symptomatic individuals.[24] WHO relies on a review of 172 observational studies showing that the use of face masks by those exposed to infected individuals was associated with a large reduction in risk of infection (up to an 85% reduced risk).[25] Still, it is not yet clear whether airborne transmission of the virus is effective (through minuscule droplets or aerosols that remain suspended in air over long distances and time). However, there are increasingly scientific arguments in favour of aerosol infection, which

---

[21] In an article published in PNAS (Proceedings of the National Academy of Sciences), the authors show the existence of a cognitive "exponential growth bias" which prevents us from apprehending the explosive growth of the Covid-19 epidemic. Their findings show the importance of statistical literacy among the general public: Lammers J, Crusius J and Gast A (2020, June 24) Correcting misperceptions of exponential coronavirus growth increases support for social distancing, PNAS, https://doi.org/10.1073/pnas.2006048117.

[22] Stadnytskyi V, Bax CE, Bax A, Anfinrud P (2020, May 13) The airborne lifetime of small speech droplets and their potential importance in SARS-CoV-2 transmission, PNAS, https://doi.org/10.1073/pnas.2006874117.

[23] Dbouk T and Drikakis D (2020, May 19) On coughing and airborne droplet transmission to humans, Physics of Fluids, https://doi.org/10.1063/5.0011960.

[24] Leung, NHL et al. (2020, April 3) Respiratory virus shedding in exhaled breath and efficacy of face masks, Nature Medecine 26, 676–680 (2020). https://doi.org/10.1038/s41591-020-0843-2.

[25] D. K. Chu et al. (2020 June 1) Physical distancing, face masks, and eye protection to prevent person-to-person transmission of SARS-CoV-2 and COVID-19: a systematic review and meta-analysis, The Lancet 395: 1973–87, https://www.thelancet.com/journals/lancet/article/PIIS0140-6736(20)31142-9/fulltext.

would explain in particular that people living in adjacent rooms (for example in quarantine hotels) can get contaminated without no direct contact[26].

These biological and physical characteristics explain the dangerousness of SARS-CoV-2. The virus has a relatively long incubation period, which means that a patient who has contracted it can be contagious without knowing it, for several days and sometimes even up to two weeks or more. For comparison, the incubation period for influenza is one to two days. Furthermore, many infected patients have no symptoms of Covid-19 while being contagious. SARS-CoV-2 is also a little more contagious than the seasonal flu, whose R_0 is of the order of 2. Last but not least, Covid-19 can be fatal (between 1 to 15% of deaths on average for confirmed cases against 0.1% for influenza, thus a case fatality rate ten to one hundred and fifty times higher).

All these data have important consequences for health authorities, and for us. First, the prevalence of mild, flu-like symptoms means that the governments and the public should redouble their efforts to promote safety precautions because there will be a natural tendency towards complacency. It's harder to mount a consistent public-health message when the virus itself is so inconsistent in whom it infects, harms and kills. However, the fact that a larger fraction of individuals survive serious infection means that health services should prepare for long-term care, and not just build up refrigerated morgues.

Second, the fact that infected people can transmit it before they develop symptoms means that detection is difficult, so widespread voluntary self-isolation and mandatory quarantine make sense. Close population monitoring and follow-up are required, with more testing and rigorous contact tracing.

Third, asymptomatic carriers are a serious issue. A huge share of infected people shows no symptoms but are contagious. A consequence is that testing strategies appropriate for other diseases, where people come to medical attention when they feel ill and are infectious, are unsuited for Covid-19. Since we cannot rely on symptoms to identify cases, testing needs to be widespread and the results returned rapidly if not immediately.

Like all coronaviruses, SARS-CoV-2 is very vulnerable to soap, hydroalcoholic solutions and disinfectants because its protective envelope, a single layer of lipids (fats), instantly dissolves in these solutions. It does not appear to be seasonal, like the flu virus, and is resistant to summer temperatures. As

---

[26]Greenhalgh T et al. (2021 April 15) Ten scientific reasons in support of airborne transmission of SARS-CoV-2, The Lancet, https://www.thelancet.com/journals/lancet/article/PIIS0140-673 6(21)00869-2/fulltext.

is often the case, the disease strikes the most vulnerable and the most socio-economically disadvantaged such as isolated elderly people, those suffering from chronic diseases including diabetes, cardiovascular or respiratory disorders, and, more generally, certain minorities and developing countries. How to get immunity to SARS-CoV-2 is still an open question, although recent works show that infected people can develop specific antibodies, particularly against the S protein, as well as a cellular immune response through the T[27] and B[28] lymphocytes, giving them protection for at least eight months after infection. Other research suggests that immunity to the coronavirus may last years, if not decades.[29] This is very good news as it may suggest that vaccines are also giving long-term protection against the virus. By comparison, SARS-CoV-1 induces post-infection immunity for about two years, after which the antibody concentration decreases. As of today, several vaccines are used across the world and several dozen vaccine candidates are still in development or in clinical trials. However, we still do not know, given that only a short amount of time has passed, whether these vaccines will provide long-lasting protection. And it will take several months if not several years to achieve immunisation of the world's population, assuming that we do not repeat the mistake made during the H1N1 flu epidemic, which saw only a majority of rich countries benefit from the vaccine. On January 11, 2021, WHO's chief scientist Soumya Swaminathan said that herd immunity to the coronavirus would not be achieved in 2021, despite the growing availability of vaccines.

In short, despite the progress of knowledge, SARS-CoV-2 is still a mystery. And virologists are still speculating about its future. Will the virus go away, like SARS-CoV-1? Will it continue to mutate, as coronaviruses are known to easily recombine? Will Covid-19 become endemic, like AIDS? According to recent works, the SARS-CoV-2 genome appears to mutate quite slowly, ten times less than HIV and twenty times less than the influenza virus. Researchers working at the Institut Pasteur have shown that the most recent sequences are separated from "sequence zero" by at most 30 nucleotide mutations and 15 amino acid mutations.[30] In a study published on August 25,

[27]Ni L et al. (2020, May 3) Detection of SARS-CoV-2-specific humoral and cellular immunity in Covid-19 convalescent individuals, Immunity, https://doi.org/10.1016/j.immuni.2020.04.023.

[28]Hartley GE et al. (2020, December 22) Rapid generation of durable B cell memory to SARS-CoV-2 spike and nucleocapsid proteins in Covid-19 and convalescence, Science Immunology, Vol. 5, Issue 54, https://immunology.sciencemag.org/content/5/54/eabf8891.

[29]Dan JM et al. (2021, February 5) Immunological memory to SARS-CoV-2 assessed for up to 8 months after infection, Science; 371(6529):eabf4063,
  https://science.sciencemag.org/content/371/6529/eabf4063/tab-pdf

[30]Zhukova A et al. (2020, November 24) Origin, evolution and global spread of SARS-CoV-2, Comptes Rendus. Biologies, https://doi.org/10.5802/crbiol.29

2020,[31] a team of doctors and researchers in Hong Kong led by Kwok-Yung Yuen describe the case of a 33-year-old Hong Kong man who was found to have two Covid-19 episodes from two distinct variants of SARS-CoV-2. The antibodies initially developed by the first strain appear to be insufficient or ineffective in preventing further contamination by the second. The possibility of reinfections is yet another reason why the health authorities, the scientific community and governments all around the world are mobilised against SARS-CoV-2.

One last point: this epidemic is anything but exceptional. The 2002-2004 SARS outbreak has many similarities with Covid-19: appearance in China of an atypical respiratory syndrome, transmission of a new coronavirus to humans (via civet) and contagion in a few months in about thirty countries. But after creating the first great fear of the twenty-first century, SARS-CoV-1 disappeared as quickly as it appeared and has not manifested since. The Covid-19 epidemic is in any case more deadly than the Hong Kong flu, caused by the H3N2 virus and responsible, in 1968, for one million deaths worldwide. Infectious diseases are the cause of nearly fifteen million deaths each year in the world and 90% of the current viruses and bacteria were still unknown in the 1980s. "Visiting past infections," Patrice Debré and Jean-Paul Gonzalez wrote, "is a lesson of humility and wisdom, and urges us to anticipate the next ones. Only the weapons have changed. The stone wall that [used] to block contagion now has a name: scientific research. Research and monitoring. While education is preparing the future inside threatened cities, the walls of science no longer aim to isolate but to promote team cooperation and the circulation of knowledge.[32]"

---

[31] Kai-Wang To K et al. (2020, August 25) Coronavirus Disease 2019 (Covid-19) Re-infection by a Phylogenetically Distinct Severe Acute Respiratory Syndrome Coronavirus 2 Strain Confirmed by Whole Genome Sequencing, Clinical Infectious Diseases, https://doi.org/10.1093/cid/ciaa1275

[32] Debré P and Gonzalez JP (2013) Vie et mort des épidémies, Odile Jacob, Paris.

2

Covid-19 Diary

Thought is an infection. In the case of certain thoughts, it becomes an epidemic.
Wallace Stevens

Since the first official infection in China on December 8, 2019, SARS-CoV-2 has grown exponentially and relentlessly, causing more than 200 million confirmed cases in 196 countries and territories and close to twenty thousand deaths a day worldwide at the peak of the pandemic (resulting in a total of more than 4 million confirmed deaths).

I retrace in this chapter the main stages of this global contagion. The goal is not to relive this painful period, as many of us have experienced health problems, lost a loved one or suffered from financial difficulties, but rather to reconstruct the history and consequences of the Covid-19 "waves" that have swept over the world.

To achieve this historical reconstruction, I relied on the excellent website of Johns Hopkins University, located in the State of Maryland in the United States, which is now the world reference for data relating to the pandemic and for its interactive map showing in real time the number of people tested positive for Covid-19 and the number of deaths in nearly two hundred countries.[1]

---

[1] https://www.arcgis.com/apps/opsdashboard/index.html#/bda7594740fd40299423467b48e9ecf6.

© The Author(s), under exclusive license to Springer Nature
Switzerland AG 2021
M. Claessens, *The Science and Politics of Covid-19,*
https://doi.org/10.1007/978-3-030-77864-4_2

When it comes to the epidemic data, there are at least three things to keep in mind. First, the data is valid at a time "t". Unless otherwise indicated, the data cited here refers to June 30, 2021, when this book was sent to print. Second, by "infected (or contaminated) person" I mean any person who carries the coronavirus, even if he or she is apparently healthy. Third, there are big differences between countries when estimating the numbers of confirmed cases and deaths caused by Covid-19. For example, during the first wave, most European countries only performed tests on patients with severe symptoms or people at risk such as caregivers. In Spain, many patients were simply followed-up by telephone given the limited testing capacities. South Korea, Taiwan and Germany, on the other hand, carried out massive screening campaigns, with some success as these three countries have relatively low death tolls. In addition, the data is not collected, compiled and corrected in the same way everywhere. More decisively, the number of confirmed cases depends on the number of tests carried out. If you don't test, you won't see anything to worry about. I still do not understand why the national authorities are not "normalising" these figures and communicating on the *positivity rates* i.e., the percentage of people who test positive compared to the total number of people who have been tested. Finally, the majority of the tests used today (called PCR, Polymerase Chain Reaction) detect the SARS-CoV-2 genome or fragments thereof. These tests are very sensitive as they can detect very small fragments of the virus by amplifying their DNA millions of times. However, a fragment is not the entire virus, and is not capable of replicating itself and infecting other human beings. Unless you know the details of the protocol, the results of the PCR do not provide information on the amount of virus present or the date of infection and cannot be used to determine whether the patient is still contagious. Last but not least, the results depend on the number of amplifications, which is generally not communicated to the public. There are two other types of tests: antigen tests, that detect specific proteins from the virus, and antibody tests, that look for antibodies that are made by your immune system in response to the coronavirus.

This practical reality makes any reliable comparison between countries virtually impossible. The same is true for the number of deaths: in France, for example, only deaths in hospital were recorded up until April 2, 2020; in the United Kingdom, deaths in nursing homes were not counted until April 28, 2020. In addition, some countries are doing everything to minimise the impact on their soil of the pandemic for political reasons and WHO guidelines have not been applied everywhere.[2] For all these reasons the official Covid-19 mortality rates have been largely underestimated.[3] One way to get around this issue is to use an interesting counting method which establishes the number of excess deaths in a given country or region during a period of time. Analysing this excess mortality shows that the total number of deaths actually caused by Covid-19 is likely to be twice as high as the official figure.[4] Other analyses suggest that the number of people who have died from Covid-19 is more than three times the recorded number.[5] On December 29, 2020, the Russian government announced, through its state statistical agency Rosstat, new figures indicating that the real death toll from Covid-19 is more than three times as high as officially reported. With these opening remarks in mind, we should not forget that most of the data relating to the impact of the epidemic should be taken *cum grano salis*—with a grain of salt.

Thus, our story begins at the end of 2019 in China, more precisely in Hubei, also called the "province with one thousand lakes," which is one of the main car industry hubs in the Middle Kingdom. It is in this central region of the country, populated by 59 million people that, according to a confidential report, the local authorities detected, on November 17, 2019, several suspected cases of pneumonia. Apparently, it took several weeks before Chinese authorities understood or recognised that this was indeed a new form of viral disease. The first announcement was made on December 8, 2019, when the Central Hospital in Wuhan officially reported the first diagnosed case of a pneumonia of unknown origin.

Three weeks later, on December 31, 2019, the Chinese health authorities issue a report on several cases of a mysterious pneumonia detected in Wuhan, immediately spotted by WHO China Country Office and WHO surveillance system. The next day, New Year's Day, WHO asks China to provide more

---

[2]As of April 16, 2020, WHO requires that "Covid-19 should be recorded on the medical certificate of cause of death for ALL decedents where the disease caused, or is assumed to have caused, or contributed to death." (WHO, International Guidelines for Certification and Classification (Coding) of Covid-19 As Cause of Death, https://www.who.int/classifications/icd/Guidelines_Cause_of_Death_Covid-19.pdf?ua=1).

[3]Only Belgium, which has one of the highest Covid-19 mortality rates with 25 000+ deaths for 11.5 million inhabitants, has consolidated all the deaths since the start of the epidemic from hospitals and nursing homes. A methodology which is controversial even for some experts, but which is vigorously defended by the Belgian health authorities.

[4]Wu J and McCann A (2020, April 21) 25 000 Missing Deaths: Tracking the True Toll of the Coronavirus Crisis, The New York Times, https://www.nytimes.com/interactive/2020/04/21/world/coronavirus-missing-deaths.html?referringSource=articleShare.

[5]The Economist (2021, May 15) There have been 7m-13m excess deaths worldwide during the pandemic, https://www.economist.com/briefing/2021/05/15/there-have-been-7m-13m-excess-deaths-worldwide-during-the-pandemic?utm_campaign=the-economist-today&utm_medium=newsletter&utm_source=salesforce-marketing-cloud&utm_term=2021-05-13&utm_content=article-link-1&etear=nl_today_1.

information and initiates an internal procedure to coordinate the follow-up and the response of its departments and regional offices. As of January 3, 2020, the Chinese authorities report that several dozens of patients with pneumonia of unknown cause have been isolated and are receiving treatment in medical institutions. The clinical symptoms are mainly fever, with a few patients having difficulty breathing, and chest radiographs showing invasive lesions on the lungs. On January 4, the organisation issues a first tweet about a "cluster" of cases in Wuhan and, on January 5, recommends all its member states to take precautions to reduce the risk of acute respiratory infections, as WHO experts believe that human-to-human transmission is possible (which China had not yet established at that time). At this point, there is still no death from Covid-19.

On January 7, 2020, the Chinese authorities confirm that it is indeed a new virus belonging to the coronavirus family and related to that of SARS. The virus does not yet bear the name SARS-CoV-2 but this day can be considered to be the official birthdate of the infectious agent responsible for Covid-19. Four days later, 41 cases are confirmed in Wuhan, including seven seriously ill and one death—a patient with serious underlying health problems. In a statement published on January 12, 2020, WHO says it is "reassured of the quality of the ongoing investigations and the response measures implemented in Wuhan, and the commitment [of the Chinese authorities—author's note] to share information regularly.[6]"

However, the number of cases is increasing very rapidly in Wuhan and in Hubei province. On January 20, 2020, Nanshan Zhong confirms human-to-human transmission of SARS-CoV-2. WHO convenes a meeting of its emergency committee on January 22 to determine whether or not the outbreak constitutes a public health emergency of international concern (PHEIC), WHO's highest level of alert. But the committee members are then divided and express divergent views. Failing to reach an agreement on the matter, WHO's Director-General, the Ethiopian Tedros Adhanom Ghebreyesus, decides to continue the meeting on January 23 at noon. This meeting also fails to secure an agreement, with several committee members saying it is still too early to declare the emergency. Nonetheless, the committee is drafting recommendations for China and other countries.

Yet breaking news that same day, January 23, 2020, see the Chinese government invoke a total lockdown in Wuhan that evening, with all travel in and out of Wuhan banned and movement restricted inside the city. At that time, the R_0 factor of the epidemic is estimated at 2.4 in Wuhan City,

---

[6]http://origin.who.int/csr/don/12-january-2020-novel-coronavirus-china/fr/.

a value which confirms the high epidemic potential of the disease. China is therefore taking the lead and going further than WHO (which was essentially going to advise Beijing authorities to provide more information, encourage public health measures, improve surveillance and testing, especially in airports and stations, and to collaborate with WHO and its partners). Journalists following the WHO emergency meeting in Geneva on that day receive an alert from the *New York Times* on their smartphones: "China has just announced, in the middle of the night, the containment of Wuhan."

China is closing its borders and stepping up controls in order to trace the millions of people from Wuhan who have gone out to celebrate the New Year with their families. Most cities are self-isolating; although they are not officially quarantined, it becomes very difficult to get out. Residents are staying at home, just going outside for basic needs. Provinces, towns and villages are closed to foreigners. The country is paralysed. New Year's holidays are extended by ten days. It is an unprecedented crisis, created by the contagiousness of the virus and the rapid spread of the epidemic. But it's too late: cases appear in Thailand, Japan, South Korea, Europe, the United States, Canada.

At this point, everyone still hopes, including the top experts, that the epidemic will be contained. In Geneva, Tedros Adhanom Ghebreyesus welcomes the "very, very strong" measures taken by China, adding that they will "reduce" the risks of spreading the epidemic beyond its borders. But these positive words do not bring consensus as several government experts immediately suspect "Dr Tedros" of weakness or even collusion with China. This is not new but this analysis is supported, among others, by the United States, which openly disregard WHO recommendations. In Geneva, the senior management responds to these criticisms by recalling the interventionism of the United States, whose governments succeeded in blocking several initiatives of the international organisation, in particular their efforts to improve access to medicines. The Director-General is obviously aware of these criticisms. Having succeeded Margaret Chan, who was criticised for her slow reaction to the Ebola epidemic in West Africa, Dr Tedros is keen to highlight the responsiveness of his organisation.

But the pandemic is about to explode in Europe. The first cases are recorded in France on January 24, 2020 (although we know now that SARS-CoV-2 had already been circulating across France since the end of 2019, so one month before the official date—and before China's official date). However, we still do not know which country of origin these first strains came from, which then contaminated all of France.

On the same day, January 24, the first clinical description of the first 41 cases of Covid-19 is published by The Lancet.[7] What is atypical in this scientific publication is that the authors, mainly Chinese doctors, express clearly their concern: they report a serious pneumonia, sometimes fatal, which requires intensive care, and "strongly" recommend the wearing of protective masks. The article, written in English, appears less than one month after the first official alerts in a prestigious journal. Compared to SARS, China, this time, was quick!

On January 28, 2020, Dr Tedros travels to China and meets President Xi Jinping. Two days later, on January 30, WHO decides to qualify Covid-19 as a PHEIC. There is no longer any hesitation among the members of the emergency committee: now is the time for action. The declaration of a PHEIC is quite an exceptional event. Since 2009, WHO has made only six PHEIC declarations, each for an outbreak requiring a vigorous global response such as H1N1 swine flu in 2009, Zika virus in 2016 and Ebola fever, which ravaged a part of West Africa from 2014 to 2016 and the Democratic Republic of the Congo since 2018. WHO's signal is unambiguous: "The [emergency] committee believes that it is still possible to interrupt virus spread, provided that countries put in place strong measures to detect disease early, isolate and treat cases, trace contacts, and promote social distancing measures commensurate with the risk".[8] However, it will take more than a month before the first European countries react and adopt containment measures.

The pandemic spreads in Italy from January 31, 2020, when two tourists test positive in Rome. The real patient zero remains unknown to this day, but the first cases of secondary infection are detected in Codogno, Lombardy, on February 18, and the situation then quickly escalates. Noting the surge in the number of cases, the Italian government places eleven municipalities in northern Italy in isolation on February 23. It is one of the first countries to suspend all direct flights to and from China. However, the country is quickly becoming the hardest hit by Covid-19.

---

[7] Huang C et al. (2020, January 24) Clinical features of patients infected with 2019 novel coronavirus in Wuhan, The Lancet, https://www.thelancet.com/pdfs/journals/lancet/PIIS0140-6736(20)30183-5.pdf.

[8] Statement on the second meeting of the International Health Regulations (2005) Emergency Committee regarding the outbreak of novel coronavirus (2019-nCoV), https://www.who.int/news/item/30-01-2020-statement-on-the-second-meeting-of-the-international-health-regulations-(2005)-emergency-committee-regarding-the-outbreak-of-novel-coronavirus-(2019)-ncov.

Europe in Lockdown

Across France, at the end of February, the number of cases is also increasing rapidly, meaning that the disease is spreading between residents. South Korea and Iran are also affected, and show rapid increases. They will be followed by Spain, Germany, the United States, the United Kingdom and many more. The gravity of the situation is now obvious to most officials. Covid-19 is becoming a pandemic. All over the world, countries are closing their borders, governments are implementing containment measures to slow down the contagion, cities are cancelling most sport and cultural events, workplaces are closing their doors and the stock market crashes, anticipating a major slowdown of the global economy. On February 25, 2020, *Science* magazine's headline is on point: "The coronavirus seems unstoppable. What should the world do now?".[9] The article quotes Bruce Aylward, a WHO official who led an international mission to China two weeks before and invites other countries of the world to follow Beijing's example and to make their minds up quickly. "Speed is everything," concludes the article.

In Italy, on March 7, 2020, as the number of deaths jumps to 233, the President of the Council of Ministers, Giuseppe Conte, decides to quarantine all of Lombardy (16 million inhabitants), including the economic capital of the country Milan, as well as the region of Venice, northern Emilia Romagna and eastern Piedmont. Those who violate the lockdown rule face fines of between 400 and 3 000 euros.

Three days later, it is the whole of Italy which is locked down: travel is only allowed for work, health care and food shopping; all gatherings are prohibited and violators sentenced by a fine or three months in prison. Italy has the second highest number of infections at this point (behind China and ahead of Iran) and the highest number of deaths in a single country. Stock markets are collapsing further and central banks are stepping in to calm them.

On March 11, WHO converts the Covid-19 epidemic into a pandemic and recommends regular hand washing, physical distancing (avoiding in particular hugs and handshakes), cancelling large gatherings and demonstrations as well as non-essential travel.

In France, the situation is also changing exponentially. Epidemic clusters appear around Paris and in the "Grand Est," the eastern part of the country. Between January 23 and March 12, nearly three thousand infection cases

---

[9]Cohen J and Kupferschmidt K (2020, February 25) The coronavirus seems unstoppable. What should the world do now?, Science, https://www.sciencemag.org/news/2020/02/coronavirus-seems-unstoppable-what-should-world-do-now.

are declared.[10] That day, President Emmanuel Macron declares in a press conference that the country is facing "the biggest health crisis in a century," and announces the closure of schools "until further notice." Two days later, Prime Minister Edouard Philippe decides to close all "non-essential" public places.

On March 15, it is Spain's turn to impose strict containment on the entire country. The next day, on the evening of March 16, Emmanuel Macron, is the next one in Europe to announce the lockdown of the entire population of France. On March 17, a day after France, it is Belgium's turn. They will be followed in particular by five American states (California on March 19, New York, Illinois, New Jersey and Connecticut on March 20), Tunisia on March 20, Rwanda and Germany (partially) on March 22, and India (partially) on March 24. In the United Kingdom and in most of the U.S. states, the situation is very different: the British government continues to advocate a *laissez-faire* attitude and their instructions boil down to hand washing. President Donald Trump refuses to impose general lockdown. As of March 20, nearly a billion people around the world are confined at home. More than 100 countries are now affected. SARS-CoV-2 has gone *viral*, or global. And Europe is now the main outbreak of the pandemic. Asia, on the other hand, is sending better news as indicators show that China, South Korea and Japan have apparently succeeded in mitigating the epidemic.

At the end of March, Europe is still in the exponential phase. Several cities are implementing even stricter restrictions with curfews, closure of food markets, access to public transport limited for workers, etc. Authorities sanction people who, defying the prohibitions, organise private parties, public meetings and sport competitions.

At the same time, the media are relaying images showing the crowded beaches of Florida, explaining that the tourist sector wants to benefit as long as possible from the economic returns that the first beautiful days allow. The political response is rather quick and police immediately close the beaches. Slowly but surely, the management of the crisis is becoming global.

In northern Italy and in Bergamo in particular, the situation is catastrophic given the lack of beds, respirators and personal protective equipment in ICUs. The media are streaming images of overcrowded emergency rooms, exhausted doctors and nurses, and military vehicles trying to "handle" the hundreds of deaths. On March 22, the government of Giuseppe Conte decides to stop all non-essential industry. On the same day, in France, the National Assembly passes a bill allowing the establishment of a "state of health emergency." The

---

[10]https://www.worldometers.info/coronavirus/country/france/.

law provides EUR 45 billion to help companies in difficulty and finance the partial unemployment of employees. Germany bans gatherings of more than two people. Jordan imposes a full curfew for two days.

At this time, several containment measures are implemented in Asia, where a second wave of contamination is emerging in China while the number of cases is increasing in India and Thailand. WHO's Director-General calls for "urgent, aggressive actions to combat Covid-19," fearing that it could lead to the collapse of already fragile health systems. The message of Dr Poonam Khetrapal Singh, Regional Director, WHO South-East Asia Region, is crystal clear: "We need to be geared to respond to the evolving situation with the aim to stop transmission of Covid-19 at the earliest to minimise the impact of the virus that has gripped over 150 countries in a short span of time, causing substantial loss to the health of people, societies, countries and economies. Urgent and aggressive measures are the need of the hour. We need to act now.[11]"

On March 23, the French Prime Minister announces a strengthening of the lockdown rules, in particular by limiting outdoor sports activities (to one hour maximum and within a radius of one kilometre around home) and closing most markets, including outdoor ones. That day the Japanese Prime Minister announces that day the postponement of the Olympic Games in Tokyo until 2021 (and they might even be postponed further). With India's decision to go into total lockdown, close to 2.6 billion people are being confined home.

One sign that the pandemic is affecting the population on a large scale is that the virus does not spare celebrities: Prince Charles of England, Prince Albert II of Monaco, the former film producer Harvey Weinstein and Michel Barnier, Brexit negotiator for the European Commission, all test positive.

On March 25, 2020, the Chinese authorities lift entry and exit controls at the borders of Hubei province and set April 8, 2020, for quarantine release in Wuhan. This marks a progressive and careful return to "normality" as China fears the arrival of new cases of contamination from abroad and other Chinese provinces.

On March 26, two great powers, Russia and the United States, are in the spotlight. In the former country, reacting to the spread of contagion, President Putin declares that the following week (March 30, 2020) all work will cease although wages will still be maintained. However, there will not be

---

[11]WHO news release (2020, March 17) WHO calls for urgent, aggressive actions to combat Covid-19, as cases soar in South-East Asia Region, https://www.who.int/southeastasia/news/detail/17-03-2020-who-calls-for-urgent-aggressive-actions-to-combat-Covid-19-as-cases-soar-in-south-east-asia-region.

any generalised or compulsory containment measures, with the exception of the closure of parks and bars. In Moscow, where the majority of Covid-19 cases are concentrated, non-food stores will not be able to open. The review provided that day by the Russian health authorities shows 182 new cases, the largest increase in one day, for a total of 840 people infected. Yet the Johns Hopkins University's website, which reports daily on the pandemic evolution worldwide, still mentions zero deaths in Russia… The situation is also taking a worrying turn in the United States, which reports more than one hundred thousand infected people and is now at the top of the list of infected countries, ahead of China and Italy. New York is becoming the epicentre of the pandemic and anticipates an explosion of hospital admissions in ICUs.

Also, on March 26, the French National Institute of Statistics and Economic Studies (INSEE) reports that the loss of economic activity is estimated at 35% compared to normal, but with large differences between the sectors of activity: a decrease of 72% is observed for car sales and 95% for some airlines companies. In contrast, several pharmaceutical, digital and media companies saw their share price soars.

In Germany, health authorities now perform nearly half a million tests per week and keep the case fatality rate at one of the lowest levels in the world (0.5%, compared to 5% in France and 7% in Spain on the same date) despite being fifth on the world list of confirmed cases. China does not record any new case of local contamination on that day even though health authorities identified 67 new imported cases. In the United States, the death toll from the coronavirus jumps to more than a thousand in the United States, with about one third of this number in New York alone.

Yet there is an encouraging sign: Italy announces, also on March 26, for the fourth consecutive day, a drop in the number of new infections, that is to say fifteen days after the start of lockdown. No country in the world is spared since the Vatican reports its first case on that day: an Italian prelate living in the same residence as Pope Francis, tests positive for the coronavirus and is hospitalised.

On March 27, British Prime Minister Boris Johnson also announces that he has contracted Covid-19 although he only feels "mild symptoms." He nevertheless decides to remain in quarantine, a decision that some observers consider more political than medical.

Televisions broadcast terrible images, typical of a time of war. In Italy, Spain, France, the United Kingdom, the United States, hospitals are overwhelmed and their ICUs saturated. There are now thousands of deaths per day across the whole world. Nursing staff admit to never having known such a situation, which many of them describe as a "massacre."

In Italy, on March 28, the media confirms the decline in the number of cases which has been observed for a few days. Nevertheless, the death toll continues to climb, with a record-high number of 969 deaths that day. This was expected, since the death curve follows that of contaminations with a lag of one to two weeks. "The main criterion to look at," explains Jean-Stéphane Dhersin, mathematician at CNRS and specialist in epidemic modelling, "is the number of patients who come to the hospital for screening and who are diagnosed positive.[12]" Dhersin also points out that this number is impacted only after at least two weeks of lockdown.

At a press conference on March 28, French Prime Minister Edouard Philippe declared: "I will not let anyone say that there has been a delay in the decision-making on containment." However, the facts, as we will see in the next chapters, do not back-up this statement.

In Italy, a draft scientific paper, pre-published on March 14, 2020 (but finally rejected for publication due, according to the authors, to long delays that made publication obsolete), shows that the epidemic in their country has been progressing silently since January, well before the official start date.[13] This is also what Dr Stefano Paglia, the head of the ICU at Codogno hospital, said in an interview with *La Repubblica* on March 3.[14] Codogno is the village which has been decimated by Covid-19 and where the so-called patient zero was identified by an anaesthesiologist. Stefano Paglia explains that a "strange pneumonia" was circulating in northern Italy from November. For this reason, he decided to increase the number of beds in the ICU by the end of December, anticipating an increase in the number of cases. Today, Stefano Paglia does not rule out the fact that there were some Covid-19 patients in Italy before Wuhan became the first epicentre of the current pandemic. This could also explain why the epidemic in Italy has escalated so quickly. On March 3, during the interview, the doctor remains optimistic: "The most absurd phase may have passed, but we could have a more dramatic phase in the next few days. By working intelligently, we will demonstrate that science cures." The advice of Doctor Paglia should have been disseminated in all countries of the world…

---

[12]Berrod N (2020, March 28) Coronavirus: plus de 10 000 morts en Italie après trois semaines de confinement, pourquoi un tel bilan ?, Le Parisien, https://www.leparisien.fr/societe/coronavirus-trois-semaines-apres-le-confinement-pourquoi-ce-nombre-record-de-morts-en-italie-28-03-2020-828 9816.php.

[13]Cereda D et al. (2020, March 14) The early phase of the Covid-19 outbreak in Lombardy, Italy, https://arxiv.org/ftp/arxiv/papers/2003/2003.09320.pdf.

[14]Visetti G (2020, March 3) Coronavirus, il primario di Codogno: Ore decisive, se il contagio si allarga sarà dura, La Repubblica, https://www.repubblica.it/cronaca/2020/03/03/news/_cosi_abbiamo_scovato_il_virus_ora_tre_giorni_per_la_verita_-250165356/.

I contacted Stefano Paglia in early 2021. This is what he wrote in an email to me on 20 April: "It is now clear for us that SARS-CoV-2 has been circulating in Northern Italy since the end of 2019. Even in our hospital, in the very first days of the epidemic, positive cases were found in patients already hospitalised without any contact with the so-called patient no. 1 or with health workers or medical staff who came into contact with patient zero. We should carefully analyse what went wrong to avoid committing the same mistakes again. If we had looked for Covid-19 in all pneumonia of viral origin, we would have realised it was circulating in our countries sooner than we could imagine." What should have been done to better handle the epidemic in Lombardy? Stefano Paglia has identified three main issues: "First, increase the testing capacity (only three laboratories were able to process tests and the average response time was three days), then add more beds in ICUs to pay adequate attention to other pathologies as not to cause further deaths, and, third, refrain from excessive confidence in therapies whose efficacy was not proven yet. What we should remember for the future is that reinforcing proximity and prevention medicine, using the diagnostic tools in a flexible way, and improving management and communication will allow all of us to reduce the devastating impact of such a totally unexpected scenario. But we should also accept the idea that it can happen…".

On March 29, the United States is now the country with the world's largest increase in the number of cases and deaths, which has doubled in four days. Nearly half of the deaths and cases are in New York state, which President Trump had considered placing in lockdown before eventually giving up on that idea. A field hospital is being built in Central Park to cope with the influx of patients expected over the next few days. Russia, the last major country to have yet to take any generalised containment measures, will seal off its borders on March 30. Moscow's mayor announces the quarantine of the city. China is once again closing its borders to avoid a second coronavirus wave, fuelled by "imported" cases. In Africa, restrictions on mobility and activity are difficult to implement and are causing a huge urban exodus. More than 3.4 billion people, or 43% of the planet's inhabitants, are confined in some 80 countries.

March 30, 2020 is another very dark day with record-high numbers of deaths announced in Italy (812), Spain (913), France (418) and the United States (573). Italy still holds the absolute world record for deaths (11 591) but also the highest case fatality rate (11% of the 101 739 cases confirmed), followed by Spain. How can we interpret these sombre figures? Guy-André Pelouze, doctor at Saint Jean Hospital in Perpignan, reminds us that interpreting this data is not straightforward: "We know that during an epidemic, vulnerable people are those most affected. There are disparities depending on the countries. Italy and Spain have a much higher death rate and death toll than Germany for those over 70. However, the age pyramid is similar in Germany and Italy. Moreover, when it comes to public health indicators, Italy and Spain are not at all badly placed.[15] Is this a difference in the performance of the health care system? Or other causes related to this one? It is difficult to say today. Crude mortality rates conceal many mysteries.[16]"

The fact remains that if the number of cases in a given country is largely underestimated, for one reason or another the case fatality rate will be overestimated in the same proportions. If we take an infection fatality rate of one percent (see footnote 18 in the previous chapter), this means that the number of infected cases in Italy is at that time ten times higher than the official figure, i.e., more than one million. This underestimation probably applies to all countries, with the possible exception of South Korea, Taiwan, Germany and, in a later stage, China, which all carried out extensive testing and screening.

"The Worst Crisis Since World War II"

Across the Atlantic, New York confirms it has become the epicentre of the outbreak in the United States, with more than 36 000 cases and 790 deaths on March 30, 2020. A 1,000-bed hospital ship has arrived to chip away at local hospitals overcrowding. Temporary hospitals have also been set up in several conference centres and under tents, notably in Central Park. But President Trump does not shy away from his optimism, announcing the recovery by June 1, 2020 (despite predicting a week earlier that it would happen by Easter). The New York Stock Exchange is up three percent at the close.

More than 3.9 billion people, or half of the world's population, are now being called on to remain in their homes to combat Covid-19. In most of the countries that have imposed lockdown, cities look the same: closed shops, deserted tourist sites, transport at a standstill, services unavailable—or overwhelmed, such as funeral directors. Yet in China, life is slowly going back to normal. Residents are gradually returning to their jobs and economic activity

---

[15] https://worldpopulationreview.com/countries/healthiest-countries/.

[16] Pelouze GA and Coursière H (2020, March 29) Covid-19: la stratégie sanitaire française est-elle efficace? Analyse comparée des résultats par pays, Atlantico, https://www.atlantico.fr/decryptage/3588395/Covid-19--la-strategie-sanitaire-francaise-est-elle-efficace--analyse-comparee-des-resultats-par-pays-guy-andre-pelouze-hugo-coursiere.

is picking up again, reaching around 70% of its normal level. In Shanghai, restaurants are reopening and customers arrive without wearing a mask. In Europe, the contrast is obvious: streets are almost empty and airports have closed their doors. The lockdowns are triggering a sort of *deglobalisation*. Most residents enjoy the exceptional air quality and the absence of noise pollution.

On March 31, 2020, nearly a million people are infected worldwide—although this is almost an underestimation as we have seen. There are probably, at the very least, five to ten million cases on the five continents, where the hospital systems of the most affected countries are struggling to cope with patient admissions. United Nations' Secretary-General António Guterres says that "We are facing a global health crisis unlike any in the 75-year history of the United Nations.[17]" More worryingly, he stresses that "we are still very far from where we need to be to effectively fight Covid-19 world-wide and to be able to tackle the negative impacts on the global economy and global societies." A few hours later, Donald Trump adopts a very serious tone for the first time, warning, at the coronavirus task force daily press conference, of a "very, very painful two weeks" ahead for the country as the virus spreads. The President presents the projections of American epidemiologists who, on the basis of Chinese and European data, predict the figure of 200 000 deaths from Covid-19 in the United States in the coming months (which will turn out to be a considerable underestimate). The situation seems to be getting out of hand, with the death toll doubling in three days. The data shows in any case that the virus has been circulating throughout the U.S. for several weeks if not several months.

On April 2, Johns Hopkins University's website announces a total number of 48 320 deaths from Covid-19 worldwide, a doubling in one week. During a virtual press conference from Geneva, WHO Director-General says he is "deeply concerned about the rapid escalation and global spread of infections." Fifty countries have entered full lockdown. We see again those surreal images, which would have been unthinkable a few weeks before: sleepy towns, lifeless streets, stores with shutters down and a silence more impressive than a Sunday dawn. From the Eiffel Tower to the Forbidden City and the Taj Mahal, most tourist sites are closed. With a reduction in passenger traffic reaching nearly 98% worldwide, most aircrafts are grounded.

On April 3, the United States continues to dominate the news. Unusually cautious, Donald Trump recommends that Americans leave their homes with their faces covered with a mask or scarf, adding, not very insightfully, that

---

[17]The AP (31 March 2020) UN Chief Says Covid-19 Is Worst Crisis Since World War II, The New York Times, https://apnews.com/article/dd1b9502802f03f88d56c34f7d95270c.

"personally, [he] will not wear one." Without saying it explicitly or without being aware of it, the American President seems to be engaged in an experiment of collective immunity (or what some call social darwinism), at least in the four states—Arkansas, North Dakota, Iowa and Nebraska—still reluctant to lockdown. Even in New York, still the most affected city in the United States, "shelter in place" remains theoretical: no fine for infringement and no control. The lax attitude of the Americans makes Europe look like they should be lauded for their decisive action! The news is also bad for Africa, which is preparing for the worst. Nigerian President Mahamadou Issoufou calls for a "Marshall Plan" for the continent. In France, Jean-Michel Blanquer, Minister of National Education, Youth and Sports, announces that the annual school year tests, the famous "baccalauréat" and the national "diplôme du brevet," will be replaced by a continuous evaluation of the students over the first three terms of the school year.

As of April 4, 2020, countries that have delayed or failed to impose lockdown display a worrying outlook. This is particularly true for the United Kingdom, the United States, the Netherlands and Sweden, which are all seeing the number of cases and deaths skyrocket. And still no deaths in Russia, according to the Johns Hopkins University website, although unofficial sources report several thousand infected cases. It is clear that not every country has the same level of transparency. On that Saturday, Donald Trump warns that America's "toughest week" of the coronavirus crisis is coming up, as the media and members of the Democratic Party lament the lack of a national strategy. Boris Johnson is hospitalised and England's Queen Elizabeth II sends a televised message to his compatriots that evening thanking health workers and calling for unity around the nation. A much-appreciated and effective initiative which helps to unite the British people against the threat of Covid-19 and gives a boost to a weakened country that could now be referred to as the "re-United Kingdom" …

At the start of the next week, on April 6, Austria announces the reopening of small shops within one week and of schools, hotels and restaurants in mid-May. Strict conditions will however still be imposed and the wearing of a mask will be mandatory.

On April 7, 2020, modelling Covid-19 scenarios for pandemic developed by researchers at the Institute for Health Metrics and Evaluation at the University of Washington in Seattle predict a heavy toll: around 150 000 deaths in Europe, including nearly half in the United Kingdom, and around 75 000 in the United States (a forecast which will be exceeded barely a month later, on May 8).

April 8 marks the end of the quarantine in Wuhan. But the return to normality is taking place gradually and there are still many restrictions in place. The majority of schools, shops and cinemas are still closed. Thanks to a smartphone app, residents know if they are in a risk zone or if they have been in contact with an infected person. The city is intact, but the damage is considerable. The economic activity has reduced by 40% according to the national news agency Xinhua and many residents are jobless, have lost relatives or suffer from psychological problems due to the very strict restrictions, the fear of catching the disease or the perspective of a second wave of contamination. In Europe, the situation is slow to improve. Italy aside, no other country has yet reached the peak of its epidemic. However, lockdown exit is on everyone's lips. Businesses are ready to go. I fear the worst: that this crisis has not taught us a lesson. In China, health authorities report 62 new confirmed cases of contamination—the fear of a second wave is very present. By contrast, the United States record nearly two thousand deaths in one day—a world record. The April 6 front page of the *New York Times* is terrifying: a map of the country shows the death toll for the big cities and a huge peak on New York with this caption: "The first 5 000 deaths came in just over a month; in less than five days the next 5 000 followed.[18]" The situation is indeed very serious in the Big Apple, with several hospitals running out of reanimation equipment, drugs, protective equipment etc. However, eight out of fifty U.S. states are still not asking people to stay at home.

A major health crisis, the Covid-19 pandemic is coupled with an economic and social crisis of phenomenal magnitude. The Director-General of the World Trade Organization (WTO), Roberto Azevêdo, says on April 8 that the pandemic "may well be the deepest economic downturn of our lifetimes" but also that "a rapid, vigorous rebound is possible." According to the International Labour Organization (ILO), 1.25 billion workers will be directly affected.

On April 9, the finance ministers of the European Union reach an agreement (by videoconference) on a common economic response to the coronavirus which will materialise through the establishment of a support plan of EUR 500 billion. That evening British Prime Minister Boris Johnson is admitted to St Thomas Hospital's ICU in London while the United Kingdom sees a death toll of a thousand per day (an underestimate since this country does not yet count deaths in nursing homes or at home). Sir Jeremy Farrar, director of the Wellcome Trust, a prestigious charitable foundation providing funding to medical research, says on television that his country is

---

[18]See the interactive map: Gamio L and Yourisholl K (2020, April 6) Toll Grew Across the U.S., https://www.nytimes.com/interactive/2020/04/06/us/coronavirus-deaths-united-states.html.

at risk of being "the most affected of European countries" by the pandemic. He wasn't wrong.

The next day, three European countries—Italy, Spain and France—show signs of a slowdown of the epidemic. Debates immediately begin to abound on the lifting of restrictions and life after lockdown. Globally, the 100 000-deaths mark has been exceeded. The New York mayor warns that schools will not reopen until September 2020.

Three days later, Emmanuel Macron announces, in a TV statement at 8 pm watched by 37 million spectators—an absolute record for French television—that the lockdown is extended until May 11, 2020, the date which has been set for the partial resumption of economic activity and the gradual reopening of schools. Wearing a mask is encouraged, but not compulsory.

In many countries, people are becoming more and more impatient and the pressure on governments is mounting to relax protective measures and boost industrial production. On that day, China records its worst toll in a month, with 108 new cases—more than double from two days earlier. The vast majority might be "imported cases," that is to say travellers returning from abroad, according to official information. This increase immediately raises concerns about a second wave, which governments hope to avoid at all costs.

On April 14, 2020, Austria and Denmark are gradually coming out of lockdown, with the opening of small businesses for the former and nurseries, kindergarten and primary schools for the latter. Spain also allows a slow return to work, on condition that masks be distributed on a large scale are worn.

On April 15, the finance ministers and central bankers of the G20 meet virtually and agree to temporarily suspend debt service payments from the poorest nations. Germany in turn announces the resumption of activities and the reopening of schools on May 4. As European Union Member States deconfine themselves in a disorganised manner, the European Commission calls on governments to coordinate their exit programmes to prevent negative effects and political tensions within the Union. Ursula von der Leyen publishes a roadmap to guide countries in this direction. That same day, Donald Trump announces he is suspending the American contribution to WHO, accusing the organisation of having made many "mistakes" on Covid-19 and of being too complacent with China. Several countries, including France, have a mixed opinion on the performance of the international organisation in the management of the health crisis, of which they criticise the slow reaction at the start of the pandemic and the lack of focus. But the American President goes further by affirming that WHO transmitted "false information on the transmission and the mortality" of Covid-19. "Today, I am halting funding to WHO while a study is conducted to examine its role in severely mismanaging and covering up the spread of coronavirus, the U.S. President declares.[19]" A position that is not very credible, as we will see in the next chapter, given the many delays and inconsistencies shown by the then American President in his own management of the crisis.

On April 16, 2020, Brazil's President Jair Bolsonaro (who will contract Covid-19 in early July) sacks his popular Minister of Health, Luis Henrique Mandetta, with whom he totally disagrees on the fight against the pandemic. A decision that comes at a critical time for Brazil, whose hospitals are close to saturation as the number of cases is surging. The minister's successor, Nelson Teich, a 62-year-old oncologist, will only hold his post for four weeks and resign on May 15 due to disagreements with the President, especially on containment and chloroquine treatment, which Jair Bolsonaro is praising and wants to impose on all infected people. A big surprise follows in the United States where Donald Trump blows hot and cold on the issue, encourages the governors of certain states to initiate a lifting of restrictions, believing that 20 states are in "extremely good shape" and "some could reopen even before the end of April." The facts will prove him wrong since the United States will experience a tremendous bounce of the epidemic in June 2020, and then in October 2020-January 2021, with a peak of contamination six times higher than that of April 2020. In fact, the first wave never went away and continues unabated.

On April 17, 2020, China suddenly announces 1 290 deaths for Hubei province, an increase of about 50% in just one day! This spectacular rebound of mortality leaves the whole world perplexed. Admittedly, since the end of March, Chinese social networks had shown long files of people queuing at the city's crematoriums, as residents were allowed to collect the urns of their deceased relatives. But this increase in the number of deaths from Covid-19 again casts doubt on the official figures provided by China. Are the Beijing authorities trying to catch-up on transparency as criticisms intensify internationally?

---

[19]Klein B and Hansler J (2020, April 15) Trump halts World Health Organization funding over handling of coronavirus outbreak, CNN, https://edition.cnn.com/2020/04/14/politics/donald-trump-world-health-organization-funding-coronavirus/index.html.

Delockdown or Relockdown?

In Europe, following Austria, Spain and Denmark, many countries are beginning or preparing for lockdown exit. From mid-April 2020, a first group of countries reopen their schools, with quite different approaches: complete (Sweden) and almost complete reopening (South Korea on April 6, Germany on April 15, China on April 21 except Wuhan on May 6, Austria on May 18, United Kingdom in June), partial reopening (Japan on April 6, Denmark on April 14, France on May 11, Belgium on May 18) and closure until September (Italy, Spain and Ireland). The aim is to boost social life and relaunch economic activity while avoiding a possible resurgence of the virus and preserving saturated health systems. The leaders also know they must respond to the frustrations of the general population after several weeks of lockdown, thus preventing the risks of social explosion. A huge challenge!

As the lifting of lockdown approaches, most European countries are preparing to strengthen their testing capacity and are calling on the private sector to develop more accurate and faster tests, most of them being based on the detection of virus nucleic acid or antibodies in the blood. Procedures are also defined for the isolation of infected people and their close contacts.

But the green light may turn red again: for example, Singapore decides on April 22, 2020 to re-establish strict lockdown until June after a second wave of contamination and nearly a thousand new cases in a single day. Two days previously, on April 20, WHO's Director-General had issued a chilling warning: "Trust us. The worst is yet ahead of us. Let's prevent this tragedy. It's a virus that many people still don't understand.[20]" On April 25, Germany is considering a new lockdown given the rise in the number of Covid-19 cases. According to a poll published on April 30, 49% of Germans believe that the lifting of restrictions is going too fast. Across the Atlantic, the United States exceeds the Covid-19 50 000 deaths mark—certainly an underestimate of the real number—and is by far the worst-affected country in the world. There are now more Americans victims of Covid-19 than soldiers of the Vietnam War between 1955 and 1975.

During a virtual press conference on May 1, 2020, WHO's emergency committee unsurprisingly confirms that the epidemic is still a "public health emergency of international concern." The time has come, however, for a revival in Europe, where some fifteen countries are beginning to relax their containment measures. In Beijing, the Forbidden City (which was indeed forbidden for three months!) reopens on the same day.

---

[20]The AP (2020, April 20) WHO head warns worst of virus is still ahead, Politico, https://www.politico.com/news/2020/04/20/who-head-warns-worst-of-virus-is-still-ahead-196214.

And, as we used to say, better late than never: on May 4, WHO and the European Commission launches the first international initiative in the midst of this global crisis. The "Coronavirus Global Response" is supported at the highest level, including by the Bill and Melinda Gates Foundation and other international organisations, and attracted from day one, donations up to EUR 7.4 billion from around 40 countries,[21] dozens of associations and various celebrities.[22] To date, pledges of close to sixteen billion euros have been received. These contributions are being used to accelerate the development of diagnostic and curative means and also to ensure equitable access to the tests for all populations, treatments and vaccines that will be commercialised. It is one of the few genuinely international initiatives set up to deal with Covid-19, in which Europeans have shown real authority and credibility. It is also worth noting the presence of China and the absence of the United States among the donors. We can obviously guess here that these decisions are not devoid of political intentions, as Xi Jinping hopes to restore China's image on the international scene and Donald Trump confirms his lack of interest in multilateral initiatives.

At the beginning of May 2020, the figures confirm that the epidemic is declining in Europe, apart from the United Kingdom which, on May 5, passes the mark of 30 000 deaths, and of 40 000 just one week later. It is now the second country in the world in terms of the number of Covid-19 deaths. The pandemic is also progressing elsewhere, and in particular in Africa, South America, Russia and India, where the epidemic has reached Dharavi in Bombay, the largest shantytown in Asia, where one million people are crowded together in precarious and unhygienic conditions. While the Indian government continues to downplay the outbreak, some sources put the death toll at seven times the official figure.

On May 9, the milestone of four million Covid-19 cases worldwide is exceeded. Brazil has more than 10 000 deaths and is expected to become the new global epicentre of the disease in June as, according to experts, the actual death toll is likely to be fifteen to twenty times higher. In Germany, the Land of North Rhine-Westphalia decides to lockdown again following the discovery of a source of contamination in a meat processing plant in Coesfeld,

---

[21] List of donors which joined the initiative on day-one: France, Germany, Japan, Norway, Canada, Spain, G20, Jordan, South Africa, Principality of Monaco, Turkey, Italy, Switzerland, Israel, The Netherlands, Grand Duchy from Luxembourg, Sweden, Portugal, Estonia, Croatia, U.S. Academy of Medicine, Bulgaria, Ireland, Serbia, Czech Republic, Poland, Australia, Denmark, Greece, Malta, Belgium, Wellcome Trust, Latvia, South Korea, Mexico, Kuwait, Slovenia, Lithuania, Oman, Romania, Finland, United Arab Emirates, China.

[22] https://global-response.europa.eu/index_en.

where as much as 10% of the 1 2000 employees have been infected.[23] A new outbreak also appears in Seoul after an infected young man frequented five clubs and bars during the previous weekend. Local authorities impose the closure of all nightlife in the South Korean capital to prevent a resurgence of the epidemic, which had been so far under control.

The next day, Boris Johnson announces an extension of lockdown at least until June 1, considering that the conditions are not yet met to return to "normal" life.

On May 11, 2020, Wuhan authorities take the massive decision to test ten million inhabitants (out of eleven million) following the appearance of six new cases which raised fears of a resumption of the contagion in the city. This intensive campaign, the cost of which was estimated at EUR 113 million, revealed only 300 confirmed cases, a mere 0.0003% of the population tested.

At a WHO virtual press conference in Geneva on May 13, Dr Michael Ryan, Executive Director of the organisation's Health Emergencies Programme, explained that "[SARS-CoV-2] May become just another endemic virus in our communities and this virus may never go away." He added: "HIV has not gone away but we've come to terms with the virus and we've found the therapies and we have found the prevention methods and people don't feel as scared as they did before.[24]"

On May 15, WHO announces it is considering a possible link between Covid-19 and an inflammatory syndrome that can develop in children and lead to a dysfunction of several organs and deterioration of health. This syndrome, which caused the death of a nine-year-old child on May 14 in France, has similarities to a very rare illness known as Kawasaki disease, affecting young children. In the vast majority of cases, children do not appear to be affected by Covid-19, but WHO states that "serious cases can still occur in these age groups." WHO is urging members to provide more information on such disease—of which only 345 cases had been documented at that time.

The Acropolis of Athens reopens on May 18, 2020. With only some 2 000 people infected and 150 deaths at that point, Greece has one of the best outcomes in Europe.

On May 19, the American states are all relaxing their restrictions but with great differences between them while the county's death toll starts to rise again with more than 1 500 deaths in a single day. A sharp increase in the number

---

[23] Some are speculating that markets and slaughterhouses are, in many countries, major sources of contamination. In the United States, this was explained by the fact that Donald Trump decided to award a bonus of USD 500 to the staff of meat processing plants, in order to allow them to operate seven days a week.

[24] WHO, Covid-19 Virtual Press conference, May 13, 2020, https://www.who.int/docs/default-source/coronaviruse/transcripts/who-pressconference-13may2020.pdf?sfvrsn=ee0d2cde_2.

of cases and deaths is also observed in Africa and South America. For its part, China decides to put the province of Jilin into lockdown, in the northeast of the country. Roads are blocked, schools closed, people confined at home: 110 million inhabitants are banned from leaving this province bordering Russia, which identifies 133 confirmed cases and two deaths.

At the beginning of June, the borders of the European Union begin to reopen in an uncoordinated manner, with some Member States pushing ahead of others, not wanting to miss the boat at the dawn of the tourist season, do not want to miss the boat. The European Commission wanted to restore free movement on its territory several weeks later as of June 15 and was hoping for a fully harmonised approach. However, the situation is different in each Member State: Italy is ready, Spain is pushing the deadline back to June 21 and the United Kingdom is imposing quarantine on travellers coming from abroad... The European Commission nonetheless successfully managed to sign an agreement on June 13 with the pharmaceutical group AstraZeneca to guarantee the supply of 300 million doses to all EU member countries as soon as a vaccine against Covid-19 is discovered.

Meanwhile in the United States, the controversy over wearing a mask has taken on a political dimension. The Heads of State and Government of the European Union all arrive wearing masks at a European summit held in Brussels from July 17 to 21, 2020 to decide on a recovery plan of EUR 750 billion.

The lifting of restrictions is progressing well in many countries, albeit with many outbreaks of the virus erupting here and there. A second wave hits Europe, whose countries in turn impose strict restrictions such as wearing masks outdoors, closing bars and restaurants or curfews at the level of communes, cities or even entire regions. Globally, the picture is even darker. "It took more than three months for the first million cases to be reported, whereas the last million cases were reported in just eight days," WHO's Director-General warns on June 22, 2020. The pandemic "continues to accelerate" across the world, the chief executive points out. In July, the global number of contaminations rise to record-high levels, with nearly 200 000 new confirmed cases every day, mainly in the United States and Latin America. A first case of Covid-19 is identified in North Korea on July 25 although its supreme leader, Kim Jong-un, had previously touted his country's invulnerability.

Technology of Hope

These "boom and bust" fluctuations surprise many people, given the widespread distancing measures and face-masks, which tend to become mandatory. However, this can be explained in a 'natural' and mathematical way. In many countries, at the time that measures begin to relax, the R_t factor is below 1, indicating that the epidemic is on the decline. Several weeks later, the epidemic resumes; new outbreaks appear, often as a result of private parties or family events, or in businesses, hospitals and nursing homes. How can we explain this trajectory when nothing is fundamentally changing? Actually, shortly after coming out of lockdown, collective behaviour most likely relaxes and people are less-practiced at adhering to "barrier gestures." Not radically perhaps but enough for R_t to go back, in certain places, to above 1. The geometric sequence then does the rest. Suppose that in city X, R_t is equal to 0.9 and in the neighbouring city Y to 1.1. A minimal difference between these two values, but one which will nevertheless have great effects. Starting from 100 cases in both cities, X will have only 26 new cases after 30 days while Y's number will be almost triple, exactly 299 (assuming an average serial interval of five days). This means a difference more than ten times greater between the two cities after just one month! The priority is therefore to defuse the potential bomb that exists in town Y before losing control of it. This is, in a nutshell, what happened in many countries after lockdown.

These dark prospects are further clouded by the fact that the disease, after having mainly affected developed countries, now strikes all continents, with very heavy consequences for the populations of the developing countries. The UN's World Food Programme (WFP) announces that the Covid-19 pandemic will see a quarter of a billion people suffer acute hunger by the end of the year, a number which almost doubles in one year, and will lead to a "global humanitarian catastrophe." Furthermore, in addition to these health and humanitarian crises, there are also unprecedented economic and social crises, with millions of jobs lost due to the slowdown in the global economy and the closure of many industries. In the United States alone, more than 30 million people declare themselves unemployed in one month (as of May 1, 2020). Most countries are facing recession and growing public debt. "The pandemic is a once-in-a-century health crisis, the effects of which will be felt for decades to come," this is what Dr Tedros said on July 31, 2020 at the opening of the fourth meeting of WHO's emergency committee on Covid-19, which reviewed the current situation.

On August 11, 2020, Vladimir Putin announces at a press conference that the Nikolai Gamaleya Institute, a State research centre in epidemiology and microbiology located in Moscow, succeeded in developing the first vaccine against SARS-CoV-2, called "Sputnik V." It seemingly consists of two different components, administered in two successive injections and using human adenoviruses to transport Covid-19 antigens. However, as of this date, clinical trials are still ongoing. In the absence of any detailed scientific publication proving the effectiveness of the vaccine, WHO and most medical associations remain doubtful. This apparently does not prevent some twenty countries from pre-ordering a billion doses. In parallel, an increasing number of countries are engaging in marketing operations and/or contract negotiations all around the world. On August 13, Argentina and Mexico announce the production and distribution within most of Latin America of the Covid-19 vaccine being developed by pharmaceutical giant AstraZeneca and Oxford University; on August 23, China starts administering a vaccine candidate to several categories of workers, including health workers while on September 1, the French laboratory Sanofi puts an end to phase 3 of their international clinical trials investigating the effect of its Kevzara drug on Covid-19 in the absence of any conclusive results.

On August 23, 2020, data updated by WHO reveals a slowdown in the epidemic for the first time, which can be seen across the world but in particular in the Americas: 1.7 million new cases of Covid-19 and 39 000 new deaths were reported the previous week, a decrease of 5% in cases and 12% in deaths compared to a week earlier.[25] But this respite will be short-lived. In Europe, the sharp upsurge in the number of cases in October testifies to the second wave which particularly hits Spain, France and the United Kingdom, even if the case fatality rate seems lower than it was in the spring. The authorities are still hesitating on the strategy to adopt even if, following Israel and Spain, several governments are considering another lockdown, implicitly acknowledging the failure of their Covid-19 public health strategy.

On September 22, 2020, the United States records its 200 000th Covid-19 death, which becomes the third most likely cause of death that year, just behind cardiovascular diseases and cancer. The number of infections is also on the rise. After months of denigrating the disease, Donald Trump tests positive on October 1 and is then hospitalised for three days. In total, at least 48 persons, including several of his collaborators, his wife Melania Trump and his spokesperson Kayleigh McEnany, test positive and many of these infections seem to be related to a ceremony held on September 26 in the

---

[25]WHO (2020, August 23) Weekly Epidemiological Update, https://www.who.int/docs/default-source/coronaviruse/situation-reports/20200824-weekly-epi-update.pdf?sfvrsn=806986d1_4.

Rose Garden of the White House for the nomination of Amy Coney Barrett to the Supreme Court, hence the name of "Rose Garden cluster" given to this Covid-19 outbreak.

On October 5, the CDC updates its health recommendations on their website, officially acknowledging that "airborne transmission of SARS-CoV-2 can occur under special circumstances," meaning that "infection can be spread through exposure to the virus in small droplets and particles that can hang around in the air for minutes or hours." The CDC therefore confirms the opinion of many scientists who have been arguing for months for better consideration of the risk of airborne transmission. This issue has been the subject of several communication hiccoughs from the American authorities. Indeed, on September 20, the CDC site already mentioned this risk on its page devoted to the coronavirus: "Airborne viruses, including Covid-19, are among the most contagious and easily transmitted." But the next day, the site reverted to its previous version, explaining that the text that had been posted had been a draft...

This second wave comes as the scientific community is increasingly divided on the strategy to be implemented to limit the spread of the virus. Some point out the collateral damage caused by lockdown and other restrictions, in particular on economic activity, social life and mental health; others like to stress the flaws in more relaxed policies, such as those followed in Sweden for example. Political and cultural differences also seem to give ground to the virus: while, in October, the number of contaminations starts to rise again in most European countries and in the United States, three new cases are identified on October 11 in the city of Qingdao, China. Immediately, the authorities decide to test (free of charge) the entire population—nearly nine million inhabitants—in just three days...

The situation is deemed "worrying" by Hans Kluge, WHO's regional director for Europe, during a virtual press conference on October 15, 2020. The official anticipates that mortality in January 2021 could quintuple compared to that of April 2020. The message is clear: it is still too early for European countries to drop their guard. Almost all of Europe is affected by the second wave and the number of new cases increases by 33% in one week according to WHO. Ireland is the first country to go back into lockdown on October 21: for six weeks pubs, bars and non-essential shops will be closed. Nearly everyone will work from home and cannot travel more than 5 kms from home. However, schools and factories remain open. This is followed by the Czech Republic on the 22nd, Wales on the 23rd (the British regions having a certain autonomy), France on the 30th, Belgium and Germany (partially) on November 2nd, Austria on November 3rd, the

United Kingdom on the 5th, Greece (partially) on the 7th, etc. Slovakia set out on October 31 to test its entire population—just under five million people. Around the world, many countries are also facing a new surge of the epidemic, such as the United States, Canada etc.

According to a preprint posted on October 28, 2020[26] and published later in Nature, a variant of SARS-CoV-2 called '20A.EU1', which apparently appeared in Spain in early June 2020, has become the dominant form of the virus in a dozen European countries. This variant, which differ from ancestral sequences at six or more positions including the mutation 'A222V' in the spike protein, spreads during the second wave which takes place in Europe. However, the article concludes that there is no evidence of increased transmissibility of this variant.

Good news: the global case fatality rate appears to be in steady decline, from more than seven at the beginning of May to less than three, which partly reflects the better treatment given in ICUs to patients with severe forms of the disease. However, there are also two additional phenomena to be taken into account: the increase in the number of tests (which tends, mathematically, to lower the case fatality rate) and the fact that the epidemic affects more and more young people (who tend to be less symptomatic).

On November 4, Sweden, which has followed a light-touch, anti-lockdown approach since the start of the pandemic, decides to close its universities in the face of the second wave which is gaining a footing in the country that has been relatively spared up until this point. The Swedish government had previously been hoping that they could slow down the epidemic and perhaps reach herd immunity in a softer manner.

On November 9, the American Pfizer and the German BioNTech announce their vaccine candidate was found to be "more than 90% effective" in preventing Covid-19, according to a first interim series of results from a phase 3 trial (i.e., tests carried out on human volunteers aimed at assessing how effective the drug is)—the last step before an application for registration. Almost 44 000 people received two doses of either the vaccine or a placebo (without knowing which one was given to them) three weeks apart. On November 8, there were ten times fewer people infected with the vaccine than in the control group, when they were exposed to the SARS-CoV-2 virus under conditions that should have made them infected. For comparison, the rate of efficacy of the influenza vaccine is, according to WHO, on average 60% in healthy adults aged between 18 and 64, with variations from

---

[26]E. B. Hodcroft et al., Emergence and spread of a SARS-CoV-2 variant through Europe in the summer of 2020, Medrxiv, October 28, 2020, https://www.medrxiv.org/content/10.1101/2020.10.25.20219063v1.

year to year. The success here lies more on the technology side: for the first time, a vaccine has been produced directly from the virus' nucleic acid (actually a "messenger RNA" or mRNA). Conventional vaccines are produced from inactivated viruses (as in the case of influenza), attenuated viruses (measles) or antigens, usually proteins (hepatitis B). Technically speaking, the vaccine candidate mRNA-1273 is a lipid nanoparticle–encapsulated, nucleoside-modified messenger RNA (mRNA)–based vaccine.

The technology has been developed in particular by Moderna in the United States and BioNTech in Germany. The idea behind it is brilliant and the principle, as often, quite simple—at least in theory. In a living organism, a mRNA directs the synthesis of one given protein, using the machinery of the host cells. Researchers proposed to produce a vaccine containing the mRNA of the S protein of SARS-CoV-2. If they could make this mRNA enter target cells in a human body, this would direct the synthesis of millions of copies of the S protein within the cells. These copies will then trigger a response by the body's immune system such as antibodies and lymphocytes, which will in turn reinforce the protection against Covid-19. An incredible discovery!

Thus, the S protein is used as a molecular weapon: the technology allows the body itself to produce the virus material against which it will then develop an immune response. The mRNA is either "naked," i.e., dissolved in the vaccine solution, or encapsulated in lipid nanoparticles. The big advantage of this type of vaccine is that the mRNA is translated directly in the cytosol (the liquid medium) of the cells, and does not need to have the virus' genetic material incorporated into the genome of the host cell. It is not without risk as new viruses might be created, at least in theory, by recombination of the vaccine's genetic material. Some side effects might also be caused by the nanolipid support. Another drawback is that the mRNA is a fragile molecule that must be stored at very low temperatures (between -70 °C and -20 °C) to prevent any damage and keep the vaccine stable. Therefore, an mRNA vaccine requires special precautions which make it more expensive than conventional vaccines. However, the technique offers immense possibilities for new vaccines, perhaps even against certain types of cancer.

The Pfizer/BioNTech achievement has been hailed as a fundamental and global scientific breakthrough—probably the only one of the whole pandemic. This is partly true. The technology had indeed been under development for many years. One of the first attempts to use an mRNA to develop a human vaccine is attributed to the Hungarian biochemist Katalin Karikó. Her starting point was a publication back in 1990 which showed positive results from an mRNA vaccine in mice. Karikó teamed up with Drew Weissman, who was working at the same university (Pennsylvania's School of

Medicine) and in 2005 solved one of the main technical hurdles by using modified nucleosides to get messenger RNAs into human cells without triggering an immune system response. The same year, after discovering this work, Derrick Rossi, a stem cell scientist working at Harvard, founded the company Moderna with Robert Langer, who immediately seized the development potential of such vaccines. Karikó and Weissman joined the company BioNTech RNA Pharmaceuticals. Several other companies, including Pfizer, were also working on projects developing mRNA vaccines against SARS and MERS, but the disappearance of these viruses led to the interruption of this research. This is reminiscent of the harsh reality that pharmaceutical giants have short-term objectives: without patients, funding sources are drying up and the research moves to other areas... For example, in Europe, each year more than 30 000 patients die due to microbial resistance, as the development of new antibiotics remains limited due to the top pharmaceutical companies' concerns regarding profitability. Today, the real research and innovation is no longer done by the big companies but by small start-ups, like BioNTech and Moderna. Some now say that Karikó and Weissman should receive the Nobel prize in chemistry for their breakthrough.

A few days after the result obtained by Pfizer/BioNTech, the promoters of the Russian vaccine assure that their candidate is "92% effective" and on November 16, Moderna announces that its candidate is 94.5% effective! Two days later, Pfizer and BioNTech publish an update that their vaccine is 95% effective, specifying that 162 members of the placebo group contracted the disease, against only 8 in the vaccinated group (or 5%), in the week following the second dose of the vaccine. On its side, China has started to distribute its vaccine. However impressive, this series of announcements looks like an auction: the winner will receive the highest offer. The so-called "science by press release" reminds us that the vaccine candidates represent a huge financial, political and health challenge. Obviously, scientists encouraged caution, as peer-reviewed publications about these treatments were (at that time) still lacking. Many other announcements followed. In its *communiqué*, Pfizer said protection was measured between one and two weeks after vaccination (i.e., when the patients' immunity is strongest). As of today, there is still some doubt about long-term immunity and protection against the different variants of the virus. Furthermore, vaccines prevent disease, but may not prevent transmission of the virus, and SARS-CoV-2 might still mutate further and evade protective immune responses. In any case, vaccines will most likely not eradicate the virus. It is now too well established to disappear following the vaccination campaign, which will not be 100% effective and is unlikely to reach the entire world's population. In some countries, the percentage of people who don't want to be immunised against Covid-19 exceeds a third of the population, as in the United States and France, and approaches 50% in Hungary, Poland and Russia. However, scientists estimate that the proportion of the population that must be vaccinated in order to stop the epidemic must be greater than $1-1/R_0$, or 66% (with $R_0 = 3$), assuming a vaccine efficiency of 100% and a duration of immunity greater than three years. This means that the epidemic will not end before several months or years have passed. The truth is that in order for the disease to peter out we need to halt transmission. And in order to halt transmission we need enough people to have the necessary levels of immunity. This may simply never happen.[27] This fact does not translate (yet) into long-term government strategies.

In this period of relative euphoria (which provokes a 3% rise of the Dow Jones index), the entire scientific community shares the enthusiasm of the vaccine promoters, despite the many questions that remain open. Across the world, scientists and doctors advising governments keep silent about the risks and uncertainties of these vaccine candidates. Mass vaccination campaigns are initiated with little or no scientific data at all. Such global excitement is unprecedented—and concerning.

"Pfizer's and BioNTech's vaccines are the start of the end of the pandemic[28]:" the headline of The Economist on November 9, 2020, is a good reflection of the general mood, although it is probably still a little too early to tell. Nevertheless, orders are pouring in. The European Commission confirms an agreement reached in September to obtain 300 million doses of the vaccine. President von der Leyen said: "This is the most promising vaccine so far. Once this vaccine becomes available, our aim is to deploy it quickly, everywhere in Europe." Meanwhile, countries are drafting their plans to be ready when the vaccine will arrive, either the end of 2020 or the beginning of 2021. Despite national differences, most governments recognise several priority groups for immunisation such as individuals having a significant risk of developing a severe or fatal form of Covid-19, for example because of their age or health status (diabetics, cancer patients, obesity etc.), caregivers and staff in hospitals and nursing homes, and people whose job is particularly important for the community and who are difficult to replace (typically in health administration, police, security services, education, etc.).

---

[27] C. Aschwanden, Five reasons why COVID herd immunity is probably impossible, Nature, March 18, 2021, https://www.nature.com/articles/d41586-021-00728-2.

[28] The Economist, Pfizer's and BioNTech's vaccine is the start of the end of the pandemic, November 9, 2020, https://www.economist.com/science-and-technology/2020/11/09/pfizers-and-bio ntechs-vaccine-is-the-start-of-the-end-of-the-pandemic.

In Europe, the heads of state and government of the European Union meet on November 19, 2020, by videoconference to strengthen their joint response to the pandemic. They seek to increase cooperation between their countries, particularly with regard to contact tracing, screening, quarantine and vaccines, and to move forward, according to the President of the European Council, Charles Michel, towards a "Union of tests and vaccines." To date, the Covid-19 testing methods used by most European countries provide results that are not comparable.

On November 20, an application for authorisation of the vaccine was filed in the United States by Pfizer and BioNTech. On December 8, The Lancet publishes the first peer-reviewed scientific article devoted to a Covid-19 vaccine, namely the ChAdOx1 nCoV-19 developed by AstraZeneca and the University of Oxford, which shows an average efficacy of 70%.[29] In China, nearly a million people have already received emergency inoculations of two experimental vaccines, Chinese pharmaceutical company Sinopharm has announced, but no clinical data showing their efficacy are available at that time. The United Kingdom is the first European country to approve on December 2, 2020 the BNT162b2 vaccine of Pfizer/BioNTech. The U.S. Food and Drug Administration gives its green light on December 11. The European Medicines Agency (EMA), which authorises and controls medicines in the Union, approves on December 21 the Pfizer/BioNTech vaccine and, on the same day, the European Commission gave the final green light, allowing the first vaccination campaigns to start across the Union.

On January 14, 2021, WHO's emergency committee met, two weeks ahead of schedule, to assess the global risk level of the pandemic and in particular the quick spread of new virus lineages (the so-called British and South-African variants) which are raising concerns about their potential to affect the efficacy of the recently developed vaccines. Preliminary data shows that the British variant does not cause any drop-off in vaccine efficacy and is even raising the efficacy of the Novavax one. However, several scientists call for a large-scale project aimed at creating a new vaccine that could work against all coronaviruses.[30]

[29]Voysey M et al. (2020, December 8) Safety and efficacy of the ChAdOx1 nCoV-19 vaccine (AZD1222) against SARS-CoV-2: an interim analysis of four randomised controlled trials in Brazil, South Africa, and the UK, The Lancet, https://doi.org/10.1016/S0140-6736(20)32661-1.

[30]Burton DR and Topol EJ (2021, February 8) Variant-proof vaccines—nvest now for the next pandemic, Nature 590, 386-388, https://doi.org/10.1038/d41586-021-00340-4.

On March 6, 2021, in a paper published by The Lancet,[31] researchers show that the AstraZeneca vaccine cuts virus transmission by nearly two-thirds. By swabbing a group of volunteers every week to detect signs of the virus, researchers found that the vaccine leads to a 67% reduction in positive swabs among those vaccinated. A very interesting result which shows that the vaccine not only protects you against the disease but also protect others. In addition, in Israel, which is at that time (early February 2021) the world's most vaccinated country against Covid-19, doctors are seeing a significant fall in infections and illness of people over the age of 60 (–50% and –40% respectively). And by mid-February 2021, the number of reported cases of Covid-19 globally declines for the fourth week in a row, and the number of deaths also falls for the second consecutive week. "These declines appear to be due to countries implementing public health measures more stringently," said Dr Tedros. An evolution which reflects changes in both government policies and individual behaviours in many countries. However, in March 2021, the number of confirmed cases is rising again and the number of new deaths plateau after a six week decrease. This time the surge seems to be associated to the spread of the 501Y.V2 variant (so-called "South-African"). In Brazil, mismanagement of Covid-19 and distrust of science threaten the whole world as the country is suffering a second wave far worse than the first. Its recorded death toll, averaging close to 3 000 a day, is a third of the world's total although Brazil has less than 3% of the world's people. Nonetheless, President Bolsonaro continues to promote ineffective and potentially dangerous drugs to treat the disease. Early April 2021, new Covid-19 cases globally rise for a sixth consecutive week (see Fig. 2.1). The largest increase in case incidence is observed in the South-East Asia and most notably in India. It seems that waves and recesses will follow one another for some time.

Vaccination is progressing in many countries, but more slowly than expected, and that which filled many with dread came to pass: in early February 2021, Europe, the United Kingdom and the United States put pressure on vaccine manufacturers to respect their contractual obligations and increase their production. In the EU, some governments are dissatisfied with the way different vaccines are being distributed among individual Member States. On March 15, 2021, Germany, Italy, France, Spain and Sweden decide to temporarily suspend the use of the AstraZeneca vaccine following reports

---

[31]Voysey M et al. (2021, March 6) Single-dose administration and the influence of the timing of the booster dose on immunogenicity and efficacy of ChAdOx1 nCoV-19 (AZD1222) vaccine: a pooled analysis of four randomised trials, The Lancet, 397(10,277):880, https://pubmed.ncbi.nlm.nih.gov/33617777/.

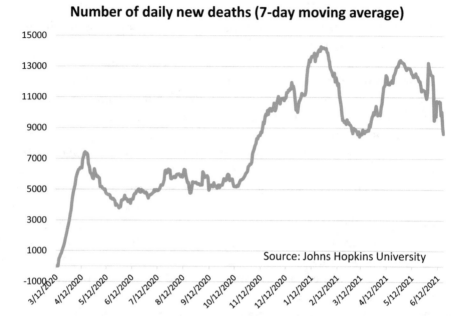

Fig. 2.1 Evolution of the number of Covid-19 deaths per day across the world (7-day moving average)

of blood clots developed by some patients. But three days later, the European Medicine Agency confirms its green light to the vaccine. On March 17, the European Commission presented a proposal that would allow EU citizens—vaccinated, tested negative or recovered from Covid-19—to travel freely across the EU-27 by the summer 2021. The plan foresees the creation of digital "green" certificates (so-called "health passports") aimed at facilitating travel from one member state to the other. On March 26, 2021, the European Affairs ministers of the European Union approved the proposal and decided to move forward "as a matter of urgency."

On May 14, 2021, Joe Biden and Kamala Harris arrive unmasked at the White House press conference and announce that masks are no longer mandatory for people who have been vaccinated. This is a remarkable step forward, but it makes even more blatant the inequalities in access to vaccines, which handicap mainly developing countries. At the same time, several countries including the United States, Germany and France, joined by the European Commission and WHO, propose to exempt Covid-19 vaccines from intellectual property rights. The aim is to facilitate and therefore accelerate the global production of vaccines for and by the poorest countries. In mid-May 2021, the vaccination rate reached 63% of the population in Israel, 54% in the United Kingdom, 47% in the United States, 29% in France but only 2.6% in Senegal and 0.8% in South Africa. This has become the next challenge: to speed up the vaccination of all mankind at the risk of repeating the tragic situations experienced in Brazil and India, among others. But international initiatives such as COVAX are underfunded. With this mixed news, which we hope will be the start of better things to come, I close this chapter on the history of the pandemic although this is of course not the end of the story. There will most probably be further developments that may require additional chapters in this book.

3

Geopolitics of the Coronavirus

I prefer liberty with danger than peace with slavery.
Jean-Jacques Rousseau

Overall, the strategies adopted by governments do not reveal major differences, with the exception of a dozen countries, which we discuss in the following pages. One thing is crystal clear: there have been incomprehensible delays and an overall failure of the decision-making process, which led to the creation of genuine "time bombs." I relate in the following pages the processes that led to this indecisiveness and in some cases the failure of governments to make any decision at all. Governments are obviously at the forefront although we have to recognise that in our democratic societies, we all have our share of responsibility.

How did governments handle the health crisis? What measures did they put in place and when? How did they work with the scientific and medical community to help them govern in a complex and frightening situation? In this chapter, we take a look at the strategies adopted by countries that have found themselves on the front line, most notably China, France, the United Kingdom and the United States. I have deliberately chosen to focus my analysis on a small number of countries as I wanted to understand in detail what happened in each of them and how the crisis was handled at the highest political level. Of course, I want to make clear that singling out these four countries does not at all mean that the situation elsewhere is any less

© The Author(s), under exclusive license to Springer Nature
Switzerland AG 2021
M. Claessens, *The Science and Politics of Covid-19*,
https://doi.org/10.1007/978-3-030-77864-4_3

worrying. The evolution of the pandemic in Africa, Latin America and some countries in Asia brings darkens an already very bleak picture. But, as you will see, focusing our analysis on how the crisis was managed in a small number of countries allows us to draw a lot of interesting lessons and hear some interesting anecdotes. There are also many broader commonalities between these countries, despite their different geopolitical situations, suggesting that the trends we will highlight must also be found elsewhere.

So why have China, France, the United Kingdom and the United States been chosen? This is a perfectly subjective decision. I wanted to consider a diverse list of some of the key countries which have played an important role in the history of the pandemic, admittedly for different reasons. These four countries have also interacted closely with the scientific and medical world but, as we will see, each in its own way. I would also like to point out that focusing on this "privileged few" does not mean that other countries will never be touched upon. On the contrary, this will not prevent us looking at what has happened and what is happening in the rest of the world. This will show in particular that after closing the borders, closing the blinds became the general rule too. Despite having faced the same enemy, few countries have learnt from each other. Cooperation has been limited and there has been little exchange on subjects as important as testing, counting methods and the best prevention and protection measures to be taken, despite repeated requests from WHO to governments to better coordinate their actions. Even the European Union has been undermined and had to struggle to rally its Member States under its 12-star banner to restore its borderless area and the coherence of its health and economic policies.

Overall, the strategies adopted by governments do not reveal major differences, with the exception of a dozen countries, which we discuss in the following pages. For countries with a high population density, the main difference is that some countries were slow to react or failed to define a coherent strategy and implement either adequate restriction measures and/or mass testing of their population, which has resulted in significant variations in terms of the number of contaminations and deaths, even if these differences are sometimes difficult to interpret.[1] Virtually all countries have experienced, at one time or another, a shortage of masks, a lack of screening tests and a shortage of resuscitation beds, with less acute problems in South Korea, Japan, in Taiwan, Germany and China in particular.

---

[1]Pelouze GA and Coursière H (2020, March 29) Covid-19: la stratégie sanitaire française est-elle efficace? Analyse comparée des résultats par pays, Atlantico, https://www.atlantico.fr/decryptage/3588395/Covid-19--la-strategie-sanitaire-francaise-est-elle-efficace--analyse-comparee-des-resultats-par-pays-guy-andre-pelouze-hugo-coursiere.

One thing is crystal clear: there have been incomprehensible delays and an overall failure of the decision-making process, which led to the creation of genuine "time bombs." I relate in the following page the processes that led to this indecisiveness and in some cases the failure of governments to make any decision at all. Governments are obviously at the forefront although we have to recognise that in our democratic societies, we all have our share of responsibility.

Yet, however disgraceful this overview is, the fact remains that the social solidarity was amazing: volunteers assembled everywhere, young people took care of the elderly, social media promoted initiatives to deliver groceries and medicines to people's front doors, restaurant chefs and families cooked meals for caregivers, to name but a few. Public and political inventiveness was remarkable. New and exceptional measures were implemented at all levels and very quickly. Some of these include, in no specific order: medicalised high-speed trains in France to transfer patients in ICUs, suspension by the European Commission of the rules for compliance with budgetary discipline, approval by the U.S. Congress to send American taxpayers up to USD 1 200, EU support for the transfer of Covid-19 patients to other countries to alleviate the burden on some hospitals, and emergency release of EUR 750 billion by the European Central Bank to support the economy.

Beijing on the Front Line

As the first country affected by the pandemic, China is also the first to have been confronted by the new coronavirus, even though recent work casts doubt on the real place of birth of SARS-CoV-2.

In early December 2019, several suspected cases of pneumonia are identified around Wuhan's Huanan seafood wholesale market. On December 16, the Central Hospital of Wuhan welcomes a worker of the market who reports having had a fever since December 1, on which usual medicines did not work at all. More people with the same symptoms are identified over the next few days. But it will take two more weeks for doctors to find out that some of these patients have been infected by a new form of coronavirus. This timeframe does not seem abnormal: when facing a new virus, doctors should first explore the most plausible hypotheses and then undertake more in-depth examinations to establish their diagnosis.

According to an unpublished government report cited by the *South China Morning Post*,[2] a Hong Kong daily, the first case of what would later be called Covid-19 is dated November 17. The patient was apparently a 55-year-old man, Jen, living in Hubei province. He could therefore be the patient zero. According to research on viral phylodynamics, scientists have been able to reconstruct the genealogical tree of the virus from the evolution of its genetic material. They estimate the date of emergence of SARS-CoV-2 (actually the "time to the most recent ancestor") back to November, 12, 2019, with a margin of error which extends from October, 11 to December, 9.[3] Did the coronavirus really first appear in China or is this simply the point at which the virus took off, initiated by the possible presence of a "super-spreader" in Wuhan at the end of 2019? We still do not know.

On December 27, another worker of the Huanan seafood market arrives at the Wuhan Central Hospital. Fortunately, doctor Fen Ai, director of the emergency department realises that the suspected patients all have SARS-like symptoms although it appears to be a different infection. In order to be certain, she asks on December 30 for a battery of additional tests for these patients: SARS, common local bacterial infections, MERS (Middle East respiratory syndrome), CT scan, etc. That afternoon she receives the results of the analyses and images she requested. By chance, the hospital respiratory department director is just passing her office at that time, and she grabs her colleague to show him the medical report she is putting together. The director is a veteran respiratory doctor in Wuhan, he fought against SARS in 2002–2003. He immediately confirms an infection by a SARS-type coronavirus. Dr Ai writes "SARS" in red on the cover page of the medical report. She immediately sends the report to some doctor friends working in Tongji Hospital in Wuhan and then posts a message on WeChat [China's most popular social media platform—author's note] for her hospital colleagues. What Dr Ai does not know at the time is that the report will quickly reach the hospital management as well as the National Health Commission in Beijing. Therefore, we retain December 30, 2019 as the official date of the first Covid-19 patient in China.

At the end of December, the situation was clear: the symptoms observed are characteristic of a new disease and a new virus for which no testing

---

[2]Ma J (2020, March 13) Coronavirus: China's first confirmed Covid-19 case traced back to November 17, South China Morning Post, https://www.scmp.com/news/china/society/article/3074991/corona virus-chinas-first-confirmed-Covid-19-case-traced-back.

[3]Nie Q et al. (2020, October 2) Phylogenetic and phylodynamic analyses of SARS-CoV-2, Virus Research, Volume 287, https://doi.org/10.1016/j.virusres.2020.198098.

method existed at the time. The Chinese authorities tried to sweep the case under the rug or, at the very least, to downplay it.[4]

Indeed, Dr Ai's message triggers a brutal reaction from local officials. First, she receives a message from the city's health authorities ordering her not to release this information to the public: "If there is panic, you will be responsible," an official tells her. Then, she is summoned by the hospital's disciplinary committee. At least eight doctors, including Dr Ai, will receive a serious warning and at least two will have to sign a letter of admonition. In reality, the matter is being handled by Beijing, where the word SARS has created the beginning of panic. Officials still remember the epidemic that started in China and spread to 37 countries, killing 774 people.

In the following days, confusion reigns in Wuhan and doctors are faced with denial by the authorities. While several dozen new suspected cases are identified, most of them are attributed to "normal" pathologies and therefore not counted as Covid-19. As of December 31, 27 patients are nevertheless hospitalised. However, according to a report from the Chinese authorities relayed by the *South China Morning Post*, there would be at least 266 infections on that day. Facing a growing number of admissions, the authorities then decide to change tack and carry out more intensive clinical investigations on patients suffering from respiratory problems in Wuhan. They then realise that a new coronavirus is spreading in the city.

That same day, December 31, in Beijing, the health authorities issue a report on several cases of pneumonia of unknown origin in Wuhan and inform WHO, which will issue its first tweet on the issue on January 4. Clearly, the Chinese authorities, who were strongly criticised at the time for the political mess and the mismanagement of SARS, want to avoid another international humiliation. However, in parallel, officials in Wuhan, including police and the Communist Party, will try to take control of what is happening and censor the doctors involved. China's leader, Xi Jinping, fails to move quickly past the early failings and shift attention to the country's drive to end the outbreak.

At the same time, Wuhan's Municipal Health Commission issues a press briefing on the pneumonia epidemic which is unusually transparent (on the part of the Chinese authorities) and provides clear messages. This is in fact the first official announcement, made on the afternoon of December 31,

---

[4]Lemaître F (2020, April 6) « Il ne faut pas diffuser cette information au public»: l'échec du système de détection chinois, Le Monde, https://www.lemonde.fr/international/article/2020/04/06/il-ne-faut-pas-diffuser-cette-information-au-public-l-echec-du-systeme-de-detection-chinois_6035704_3210.html.

which I am copying here almost in its entirety[5]: "Recently, some medical institutions in Wuhan found that many of the pneumonia patients they have diagnosed are related to the Huanan Seafood market. After receiving the report, the Municipal Health Commission immediately conducted a case search in the city's medical and health institutions and a retrospective investigation in the Huanan market. Twenty-seven cases of pneumonia have been found, of which seven are in serious condition, and the remaining cases are stable and controllable. Two patients are expected to be discharged in the near future. The clinical symptoms presented by the patients were mainly fever, a few patients had difficulty breathing, and chest radiographs showed bilateral lung infiltrative lesions. At present, all cases have been isolated for treatment, follow-up investigations and medical observations of close contacts are ongoing, and hygiene investigations and environmental sanitation of the Huanan market are also ongoing. Wuhan City has organised clinical, epidemiological, and virological expert consultations with Tongji Hospital, Provincial CDC [China's Centre for Disease Control and Prevention—author's note], Wuhan Institute of Virology, Chinese Academy of Sciences, Wuhan Infectious Diseases Hospital, and Wuhan CDC. According to the analysis of epidemiological investigations and preliminary laboratory tests, the above cases are considered to be of viral pneumonia. So far, the investigation has not found any obvious human-to-human transmission, and no medical staff infection has been detected. The detection of the pathogen and the search for the cause of the infection are in progress."

The press briefing mentions "viral pneumonia." *Caixin*, the Chinese economic news outlet which disclosed several noteworthy reports during the epidemic, affirms that several laboratories received samples from hospitals in Wuhan at the end of December, taken from patients with atypical pneumonia, and sequenced the viral genome long before its official announcement. Although they warn the authorities of the possible dangerousness of this unknown coronavirus, they were instructed on January 1 to destroy the samples. According to *Caixin*, the first signs of the emergence of a new SARS-like virus have been identified, shared and then deleted.[6]

On December 1, 2020, more than one year later, *CNN* and the *Wall Street Journal* cite a confidential report from the Hubei CDC according to which several cities in the province, including Wuhan, recorded in December 2019

---

[5]The official communiqué (in Mandarin) can be found on http://wjw.wuhan.gov.cn/front/web/showDetail/2019123108989.

[6]Yu G et al. (2020, February 29) In Depth: How Early Signs of a SARS-Like Virus Were Spotted, Spread, and Throttled, Caixin, https://www.caixinglobal.com/2020-02-29/in-depth-how-early-signs-of-a-sars-like-virus-were-spotted-spread-and-throttled-101521745.html.

a number of influenza cases up to twenty times that of the previous year.[7] While there is not necessarily a link between this influenza epidemic and Covid-19, it is troubling that Chinese authorities have not made this information public and could suggest that the number of Covid-19 cases in China is underestimated.

Medical uncertainties and censorship by the authorities contributed to a three-week period of hesitation during which victims of Covid-19 were certainly left untreated or excluded from official reports. Of course, statistical data in China is always a very political and sensitive matter and there is no doubt that Beijing exerted pressure on local officials to control the situation and resolve the emerging epidemic as quickly as possible. There is no formal proof that the official figures provided by the Chinese authorities—currently around five thousand Covid-19 deaths—have been manipulated, but this hypothesis cannot be ruled out either.[8] The political leaders of the Middle Kingdom have always been keen to instrumentalise and even rewrite their history. However, the case takes on an international element as several governments appear to be willing discontinue their bilateral cooperation agreements were the Chinese authorities unable to transparently communicate on Covid-19. During these first weeks, when the epidemic was smouldering, the media strongly criticised their sluggish reaction and the mistakes made by the Chinese government, which is accused of minimising the crisis, suppressing information and denying the emergence of a new disease. Of course, experts recognise that it takes some time to cope with a new virus, which explains why cases can go undetected. This also happened in Europe and elsewhere in the world, where doctors had not instantly acquired the reflex to think "Covid-19" when receiving patients with respiratory problems.

Let us note here that the data and statistics provided by the Chinese authorities on Covid-19 have not been called into question by WHO. Furthermore, the downward trend observed since mid-March 2020 is undoubtedly correct since the Chinese authorities have not hidden the sharp rise in April 2020. But there is no doubt that the Chinese Communist Party —and therefore the highest echelons of government—has, for several weeks, played down the facts and the risk posed by the new virus. You will also

---

[7] Walsh NP (2020, December 1) The Wuhan files, CNN, https://edition.cnn.com/2020/11/30/asia/wuhan-china-covid-intl/index.html.

[8] On December 28, 2020, Yanzhong Huang, a senior fellow of the US Council on Foreign Relations tweeted that a China's CDC nationwide survey for Covid-19 shows that Wuhan's seroprevalence rate was 4.43% in April 2020, suggesting the infection of 480,000 people, compared to 50,000 officially confirmed cases in the city at that time. However, the results of the survey seem to be available on WeChat only and accessible only to people living in China: http://china.caixin.com/2020-12-28/101644222.html.

remember that President Xi Jinping was absent from public view in the first weeks of January 2020. In a speech made on February 3, 2020, Xi said he knew about the coronavirus outbreak in early January and that he has been personally directing the response to the outbreak (while at the same time Wuhan officials were publicly downplaying its danger).

The media also commented on the role played by another doctor in Wuhan, Li Wenliang, who has often been portrayed—in my opinion wrongly—as a whistle-blower. It is Fen Ai who first sounded the alarm. It is her who alerted her colleagues and the authorities. It is also her who gave an interview on March 10, 2020 with Chinese magazine *Renwu*,[9] in which she is very explicit: "If I knew what would happen next, I would have ignored the critics from the government which instructed me to keep silent. I regret I didn't keep screaming out at the top of my voice and let as many people as possible know the truth." A study[10] found indeed that if China had moved to control the outbreak three weeks earlier, it might have prevented 95 percent of the country's cases. Trying to control the medical staff and hospitals and build a new narrative, China wasted time and energy in the first weeks of the epidemic that could have been used to slow infections in China before they exploded into a pandemic.

Thus, on December 30, Dr Wenliang, who worked as an ophthalmologist in Wuhan, receives the message from Dr Ai unveiling the presence of a SARS-type coronavirus in Wuhan. A full report had already been circulating for three days in medical circles in Wuhan. The same day at 5:43 pm, Li Wenliang posts a message on WeChat and send it to a hundred of his professional contacts, warning them that "seven SARS-like cases have been confirmed at the Huanan seafood market." The doctor urges his colleagues and friends to be careful about the risk of infection. But Li Wenliang is quite upset when he discovers that his post has been widely shared on the social network in a truncated version, with the comment: "SARS is coming back to Wuhan."

Like Dr Ai and six other Wuhan doctors, Li Wenliang just wanted to warn his contacts that a SARS-like virus had emerged in his hospital. But on January 3, along with one of his colleagues, Dr Wenliang is summoned by the police, where they are both reprimanded for spreading rumours and "untruthful speech" on the Internet.[11] The two doctors agree to sign a letter

---

[9]https://blog.dwnews.com/post-1313840.html.

[10]Lai S et al. (2020, May 4) Effect of non-pharmaceutical interventions to contain Covid-19 in China, Nature, 585, pages 410–413, https://doi.org/10.1038/s41586-020-2293-x.

[11]D. Clarke, Wuhan police "letter of reprimand" to Li Wenliang: translation and analysis, The China Collection, February 6, 2020, https://thechinacollection.org/wuhan-police-letter-reprimand-li-wenliang-translation-analysis/.

of reprimand, written by the authorities, in which they acknowledge not to spread rumours, in order to remain free as they believe their duty is to work and help the population facing the growing epidemic.

In the doctor's subsequent interview with reporters of Nan Du[12] and from the reactions of his colleagues, Li Wenliang does not appear to be a whistle-blower. In fact, he just did his job, as a doctor and teacher. His mistake was to communicate freely on a subject that had become sensitive, without going through the authorities. A week later, Doctor Li contracts Covid-19, probably after examining a patient on January 8.

A few days earlier, on January 5, the genome of the novel coronavirus is isolated and sequenced by the team of Professor Zhang Yongzhen of Shanghai Fudan University, who confirms that it is indeed a new virus. He immediately notifies the Shanghai Municipal Health Commission as well as the National Commission, prompting the CDC to send an emergency alert on January 6. On January 11, Prof. Zhang's team publishes the complete genome sequence in two public databases, Virological.org and GenBank, enabling the development of the first testing methods. The researchers also provide WHO with the full data the next day. That same day, several pages of websites devoted to the new virus by Fudan University and the Wuhan University are deleted. These pages referred in particular to the authorities' decision to check any new scientific article before publication. Also, that day, health authorities in Beijing report to WHO that no new case of Covid-19 has been detected since January 3. These events seem to show that China's central power is taking charge of the management of the epidemic, and specifically the information about its origin and what happened during the first weeks of the outbreak. More importantly, it was this centralisation of power that made the whole system fail. As interviews with doctors and officials, leaked government documents, and Chinese media investigations revealed the depth of the government's failings, the local health administration and medical expertise had no choice but to succumb to tampering by the state.

The availability of better testing techniques is welcome news at this point as they will help to provide a more accurate picture of the number of cases—until then, cases were mainly confirmed on the basis of clinical symptoms, which sometimes are wrongly attributed to other diseases. Moreover, the number of cases, which had been made public and updated daily from the beginning of January, effectively fell once the tests become available, since

---

[12]https://new.qq.com/omn/20200130/20200130A0JX7G00.html.

suspected patients may test negative. Nevertheless, life goes on and it is business as usual: on January 18, the Chinese Communist Party is still organising a gigantic banquet for 40 000 people in Wuhan!

However, the Chinese government eventually decide to take action. It is not clear if the information sent by the officials and hospitals in Wuhan played a key role in this. However, it turns out that the reports of Nanshan Zhong and of a small WHO delegation, which came to Wuhan on January 19 and 20, have been decisive. Nanshan Zhong heads the Public Health Commission in Beijing which coordinates the fight against infectious diseases at national level. On June 1, 2020, Nanshan Zhong confirmed to me that this visit and the report he and his colleagues gave to the Commission influenced the government's decision to put Wuhan into lockdown. Nanshan Zhong's scientific credibility and his proximity to political circles enabled him to obtain the best information in Wuhan (which does not always reach Beijing). He also acknowledges that the government took into account the recommendations of the Commission to isolate and create a "cordon sanitaire" around the city—"Confine Wuhan, do not go to and do not leave the city"—while advising that public health measures might be implemented elsewhere (in particular physical distancing, travel restrictions, wearing face mask, temperature checks, intensive screening, contact tracing and early isolation, extension of school holidays, cancellation or postponement of large public events and mass gatherings, and closure of libraries, museums and workplaces). Nanshan Zhong also feels that the government had been listening to the experts and in particular the executive committee of the Commission, and followed their advice for monitoring the evolution of Covid-19 in China. This was confirmed to me, also on June 1, by another scientist well known in China (and appreciated for his freedom of speech), Wenhong Zhang, professor at the Fudan University in Shanghai (Fig. 3.1).

In her "Wuhan Diary," Chinese writer Fang Fang reports, in sixty posts, about life in quarantine.[13] She explains that, up until January 20, 2020, authorities have always claimed that the disease was under control and that the virus is not transmitted by humans and describes the fear and panic that spread throughout the population on that day, following rumours of an imminent lockdown. She also wonders why, after the first information about the virus circulated from December 31, nothing happened since. "Haven't we learned anything from the SARS epidemic in 2003?," Fang writes. According to her, the answer lies in the excessive confidence of the Chinese in their

---

[13]Fang F (2020) Wuhan Diary, HarperVia, New York.

Fig. 3.1 Chinese President Xi Jinping awards the Medal of the Republic to Nanshan Zhong during a meeting in the Great Hall of the People in Beijing on September 8, 2020. (© AFP)

government, which few of them believe capable of such laxity and irresponsibility when human lives are at stake. She also points to behaviours deeply rooted in the head of politicians, such as communicating good news rather than bad, preventing people from disseminating and speaking about the truth, failing to uphold the value of the individual, etc. (By the way, let me underline the freedom of tone and the independence of Fang Fang, which is very critical of the authorities—something quite normal in the West but which is only undertaken with great risk in China.)

On January 23, three and a half weeks after the outbreak of the virus and two days before the Chinese New Year, the central government puts Wuhan into lockdown, with city officials prohibiting all transport in and out of the city. As the decision was announced overnight, residents of Wuhan did not have time to plan for a possible departure. Effective from 10 am, almost the entire population of Wuhan is locked down for more than two months. It is important to note that a Chinese lockdown in no way compares with what will happen in the West: leaving the house is only allowed for food shopping and health care, and the city government of Wuhan imposes the wearing of a respiratory mask, which most of the inhabitants have in any case started to do since the beginning of the week.

In the first days of the quarantine, the implementation of controls and health measures is frenetic. Neighbourhood committees visit homes to check the temperature of all residents. Because of the lack of tests and hospital beds, patients with mild symptoms are urged to stay at home, knowing that they risk infecting their loved ones. In some cases, residents are forcibly taken into isolation. Hundreds of infected people walk from one hospital to another to try to find a treatment (public transport has been halted). However, local organisation and logistics will quickly improve. Two hospitals are built in ten days to accommodate the number of infected people that is growing every day. Eleven (temporary) field hospitals are set up to welcome mild but confirmed cases. Stadiums, schools, hotels and convention centres are requisitioned and turned into places of isolation for people who have been in contact with confirmed patients. After a week, some eight thousand tests are being done in Wuhan daily. A military discipline is sweeping the whole city. Patients are divided into four categories: those with severe symptoms are admitted to selected hospitals for treatment of the disease; those who test positive but with mild symptoms are sent to one of the eleven provisional hospitals; those suspected of having Covid-19 are admitted to requisitioned and quarantined hotels; finally, those who have been in contact with confirmed cases are also quarantined in hotels or school dormitories.

The day after lockdown, on January 24, Li Wenliang's health rapidly deteriorates. The doctor is admitted to the respiratory medicine department of Wuhan's Central Hospital. On January 31, he relates his experience at the police station and publishes the letter of reprimand on WeChat. His post goes viral, sparking international protests against Chinese authorities. Faced with this unexpected media coverage, the Communist Party backtracks—an extremely rare event. On February 4, the Supreme People's Court declares that the eight citizens of Wuhan should not have been punished, insofar as their words were not completely wrong. The Court even posted a message on social media: "It would have been desirable for the public to have taken the 'rumours' seriously and then started wearing masks, taking sanitary measures and avoiding the seafood market." But, as usual, the Communist Party is organising a "clean-up operation" and focuses blame on local bureaucrats, including for censuring doctors who warned others about the infections. The Party's leaders promptly dismiss two health officials and, later, the Party's secretaries in Hubei Province and Wuhan. "The same thing happened in 2003, during SARS," says Jean-Philippe Béja, emeritus research director at CNRS.

On February 6, Wenliang's family and friends find out that several Chinese state media mistakenly announced Li's death. A monumental fault as the messages were quickly erased from the Internet and the networks. How could the central power at that moment have been so certain that the doctor's death was imminent? His death was indeed confirmed the next day by Wuhan Hospital. Probably the first doctor to die from Covid-19, Li Wenliang reminds us that in present-day China, public information and the media are still very tightly controlled.[14] In mid-April, *Caixin*'s deputy editor-in-chief Gao Yu explains in a podcast that only 30% to 40% of the information uncovered by his journalists, who spent the seventy-six days of confinement in Wuhan, has been published, meaning that much has been censored. According to recent reports, Chinese authorities organised a large-scale "purification" of the Internet immediately after Wenliang passed away. They instructed news websites not to issue push notifications of his death, told social platforms to remove his name from trending pages, and stimulated legions of fake online comments to flood social sites with distracting chats.[15]

After Wuhan, Hubei goes into lockdown two days later, on January 25. The whole province, about 59 million people—almost as many as in Italy—is cut off from the rest of the world. According to local residents, the population complies with the restrictions and behaves responsibly—actually typically Chinese—despite the fact that the lockdown is very strict. Infected people must leave their accommodation and stay in a specialised centre. During two months of strict quarantine, streets and sidewalks remain empty. The vast majority of Hubei's inhabitants are cloistered at home and barely dare to open their doors to the delivery workers who sometimes pass by to bring them a prepared dish. In some residences and villages, the gates are padlocked and the doors welded. On WeChat, residents despair of not seeing the sky for more than eight weeks: "We will not have seen the cherry trees blossom."

So, from the official identification of the first Covid-19 case in China, on December 30, 2019, it took three and a half weeks for the government to decide to quarantine Wuhan and then all of Hubei. This will be our "benchmark," our point of reference for comparing with other countries. One could argue that the first Covid-19 case may have appeared on November 17 in

---

[14]As I visit this country regularly, it is always unpleasant for me to lose access to the major international media and social networks during my stay.

[15]Zhong R, Mozur P, Kao J and Krolik A (2020, December 20) No Negativity: How China Censored the Coronavirus News, The New York Times, https://www.nytimes.com/2020/12/19/technology/china-coronavirus-censorship.html.

Wuhan, but in the absence of any confirmation, we will stick here to the official dates. When Wuhan went into lockdown on January 23, the Covid-19 toll has officially 830 confirmed cases and 30 deaths. The real numbers are most likely much higher. By comparison, Italy and Spain had around 800 deaths when they went into national lockdown.

In the end, and with some hindsight, many international outlets hailed the response of the Chinese government, whose reaction was, compared with other countries, firm and effective, in short "a police reaction." Three and a half weeks sounds like a very long time, but very few countries have been as quick as China to contain the epidemic. And Europe is far behind. Another point noted in the press: since the SARS outbreak, the Chinese authorities have improved their public health communication which is evidenced in particular during the pandemic of 2020—while obviously still not being completely transparent.[16] If we rely on the official data, their strategy is a success, as the mortality rate from Covid-19 in China is one of the lowest in the world.

During an electronic discussion on March 23, 2020, François Ouellette, professor at the University of Chengdu, approximately 500 km west of Wuhan, gave me some interesting insight on the Chinese epidemic: "What the Chinese government has done, and what all the other governments in the world should have done, is quite simple: they listened to the experts. What did experts like Zhong Nanshan, the man who fought the SARS epidemic, do? Well, they looked at the number of cases, plotted the curve with what little information was available, and nevertheless estimated the reproduction rate of the virus (the famous R_0 factor). Their conclusion was clear cut. Without "non-pharmaceutical interventions," millions and millions of people would have been infected within weeks. Little was known about the rate of mortality from the virus, as there were still few deaths, but it was clear that the infection rate was much higher than that of the flu. The decision to shut down not only Wuhan and Hubei, but essentially all of China, meant a huge economic loss. But it was also clear that if it had not been done, the loss would have been even greater. China then deliberately sacrificed at least two months of its economy to save the country. Today, there is no doubt that it was the best decision. The results are there: from mid-February 2020, we went from more than two thousand infections per day to a few hundred, then a few dozen at the beginning of March. It is equally clear that the governments of Europe and America which refused or were slow to follow this route,

---

[16]Frost M (2019, May 10) Progress in public health risk communication in China: lessons learned from SARS to H7N9, BMC Public Health 19:475, https://bmcpublichealth.biomedcentral.com/articles/10.1186/s12889-019-6778-1.

have paid a high price for their short-sighted view." In March 2021, one year after the Wuhan and Hubei lockdowns, several Western governments seem to recognise the merits of the Chinese approach. There is indeed an increasingly vocal group of academics and experts across Europe that are calling for governments to implement stricter measures to try and drive down Covid-19 transmission rates instead of implementing what they see as "half-measures." The aim is to reach a point where Covid-19 would be brought down as close as possible to zero through in particular the use of zoning and travel restrictions in Covid-19 hotspots, to defeat the pandemic.

Of course, the analysis of the crisis management in China raises the question of the official data's reliability. *Caixin* affirms on March 23, 2020 that many asymptomatic cases have not been taken into account in the official reports.[17] The magazine also claims that the long queues at the Wuhan's seven crematoriums around March 25, 2020 could reflect a surge in infections (although this congestion might also have been created by the closure of these establishments for ten days on New Year's Eve). According to some sources, more than 40 000 people have died in Wuhan since the start of the quarantine—ten times the official figures.

The press did not say much about the living conditions of the Chinese population during this period. Actually, the disease has taken a heavy toll among Wuhan and Hubei residents. First, law enforcement brigades put all those who were tested positive or suspected to test positive into isolation. I cannot imagine such a decision being made here in Europe or in the United States. Moreover, I have not seen the slightest discussion on the impact of such a measure, which is undoubtedly positive from an epidemiological point of view but places a heavy toll on psychological and personal health. In Europe, people who test positive are invited or requested to stay at home, with the risk of infecting their spouse, their children, their neighbours. Furthermore, the residents of Hubei and, to a lesser extent, of other Chinese provinces were facing a second difficulty unique to China: the complete lack of income following unemployment. Most of those who were forced out of work, whether for health or economic reasons, received no financial compensation at all from the public authorities. Unemployment benefit does not exist in China. Fang Fang clearly describes the very difficult living conditions in Wuhan during the lockdown: halting all public transport, controlling the increase in the prices of masks and other essential goods, frequent police checks, etc.

---

[17]Jinzhao Z and Yin D (2020, March 23) Despite Official Figures, Wuhan Continues to Find New Asymptomatic Covid-19 Cases Daily, https://www.caixinglobal.com/2020-03-23/despite-official-figures-wuhan-continues-to-find-new-asymptomatic-Covid-19-cases-daily-101532880.html.

Despite all these difficulties and shortcomings, the management of the crisis in China shows the success of a non-pharmacological strategy based on social distancing—actually the only possible approach that could have been taken in early 2020 as we did not have any effective antiviral drug or vaccine at that time. But this strategy, in China, was implemented in a quasi-military way, by physically isolating all symptomatic people and by quarantining those who had been in contact with infected people. In some European countries, quarantine has only been mandatory for international arrivals or nationals returning home from abroad.

Research published in *Science* confirms that restricting the mobility of people with the closure of borders and the confinement of populations has reduced the territorial expansion of the epidemic and hence the final number of cases.[18] Using real-time travel data provided by the Internet services company Baidu on the mobility of both infected and healthy people, the international team signing this article shows that the spread of SARS-CoV-2 in China at the start of 2020 correlates with mobility statistics. Researchers conclude that the drastic control measures implemented in China at local level such as social isolation and hygiene, rather than long-distance travel restrictions, substantially mitigated the spread of Covid-19. However, on the basis of the data collected, they recommend that overseas travel should be restricted until the pandemic is over. A more recent study, carried out in Europe, also confirms the important role of mobility, which alone explains the initial spread of the virus in Italy, France and Spain.[19] This research provides the scientific evidence that backs up lessons learned from the great plague epidemics of the Middle Ages to the nineteenth century: that isolation, quarantine and containment measures appear to be the only protective and preventive strategies in the absence of any effective treatment.[20] Such measures have also been used more recently, in particular during influenza and SARS epidemics, although we also know that the restriction of social interactions has a negative impact on the ability of society to return to normal afterwards. However, according to the epidemiologist and historian Patrice

---

[18]Kraemer MUG et al. (2020, March 25) The effect of human mobility and control measures on the Covid-19 epidemic in China, Science, https://science.sciencemag.org/content/early/2020/03/25/science.abb4218.

[19]Iacus S et al. (2020) How human mobility explains the initial spread of Covid-19, EUR 30,292 EN, Publications Office of the European Union, Luxembourg, https://ec.europa.eu/jrc/en/publication/how-human-mobility-explains-initial-spread-Covid-19.

[20]The concept of quarantine, that is to say the coercive (but transitory) isolation of people and goods suspected to be contaminated, dates back to 1377 when the rector of Ragusa, present-day Dubrovnik, in Croatia, imposed on any ship arriving from an infected area a thirty-day isolation on the small island of Mercado (Debré P and Gonzalez JP (2013) Vie et mort des épidémies, Odile Jacob, Paris).

Bourdelais, these measures have proven effective in the past in slowing the development of infectious diseases.[21]

Taking account of the political situation in China, of which we are all well aware, the present day challenges it faces, of which we are less well informed, and the fact that the epidemic started there, it is fair to say that this country has had remarkable success in its handling of the crisis. This was done in its own way of course, with errors made along the way, but also with indisputably positive results, both in terms of health and economics. The scientists and politicians I know in China always impressed me with their professionalism and commitment. World-renowned experts like Wenhong Zhang and Nanshan Zhong answered my emails overnight. Without yet knowing all the details of the development of the epidemic in this country, scientists and doctors have played a leading role in influencing if not determining the political response. There has been, as we have seen, some hesitation and attempts to secretly control or even quell the emerging epidemic. However, frightened by the SARS crisis, health officials and political authorities in Beijing took things in hand in early January. From February 16 to 24, 2020, a WHO delegation visited China to assess the response to the coronavirus and thus offer recommendations to countries not yet affected. Here is the conclusion of their mission report: "In the face of a previously unknown virus, China has rolled out perhaps the most ambitious, agile and aggressive disease containment effort in history [...] Achieving China's exceptional coverage with and adherence to these containment measures has only been possible due to the deep commitment of the Chinese people to collective action in the face of this common threat. [...] China's bold approach to contain the rapid spread of this new respiratory pathogen has changed the course of a rapidly escalating and deadly epidemic."[22]

In a letter published by Science on May 14, 2021, scientists note that "it was Chinese doctors, scientists, journalists, and citizens who shared with the world crucial information about the spread of the virus—often at great personal cost. We should show the same determination in promoting a dispassionate science-based discourse on this difficult but important issue."[23] Today, Beijing is putting forward its management of the crisis as a planetary model and offering help to any country that requests it. It is not certain that the Middle Kingdom will receive the dividends of this strategy, which

---

[21] Bourdelais P (2003) Les Epidémies terrassées: Une histoire des pays riches, La Martinière, Paris.

[22] WHO (2020, February 16–24) Report of the WHO-China Joint Mission on Coronavirus Disease 2019 (Covid-19), https://www.who.int/docs/default-source/coronaviruse/who-china-joint-mission-on-Covid-19-final-report.pdf.

[23] Bloom JD et al. (2021, May 14) Investigate the origins of COVID-19, Science, https://science.sciencemag.org/content/372/6543/694.1.

gives an impression of great generosity but which is also seen as inappropriate political arrogance and diplomatic conquest. As early as April 2020, several countries, led by the United States, criticised the Beijing government harshly for its responsibility in releasing the virus, its poor communication and the pressures on WHO. There are still many grey areas in the account given by Chinese authorities about the epidemic and what happened in Wuhan. And the government is still actively involved in rewriting the early history of the epidemic. It took several months for the Chinese government to allow an international delegation of scientists from WHO and ten institutions and countries to travel to China in early 2021. The aim of this politically sensitive mission was to engage in and review scientific research with their Chinese counterparts on the origins of the virus. Added to this Covid-19 context are more political calculations: many countries take a dim view of China's leading role in the post-COVID world. We will see in the following pages that France, the United Kingdom and the United States have little to teach us in terms of crisis management. But the mistakes China made in the first few weeks, the pressures on the medical community and the opacity of its communication will continue to weigh on global diplomatic relations. The Chinese are the first to criticise the missteps of their President Xi Jinping. But only in private.

In Paris, a Plethora of Scientific Committees

Unsurprisingly, the management of the health crisis in France is very different from China, although communication has been no more transparent.

The first three Covid-19 cases in France, incidentally also the first confirmed cases in Europe, are officially recorded on January 24, 2020. These are three Chinese nationals who had stayed in Wuhan. They all suffer from fever and coughing. Two of them are hospitalised in Paris, the third in Bordeaux. These are the "official" cases, which will trigger the authorities to take action. However, according to a scientific article published on May 3, 2020, Covid-19 was already present on French territory at the end of 2019[24]. Indeed, doctors of the Paris Seine Saint-Denis hospitals decided in March 2020 to re-examine the files and respiratory tracts' samples of patients who had been admitted between December 2, 2019 and January 16, 2020 for influenza-like symptoms (temperature above 38.5 °C, cough, rhinitis, sore

---

[24]Deslandes A et al. (2020, May 3) SARS-COV-2 was already spreading in France in late December 2019, International Journal of Antimicrobial Agents, Volume 55, Issue 6, https://doi.org/10.1016/j.ijantimicag.2020.106006.

throat). Fourteen suspected cases were then tested for PCR, and one tested positive for SARS-CoV-2. The case is a 42-year-old patient who was hospitalised on December 27 in ICU for haemoptysis, i.e., coughing up of blood. Other French hospitals launched similar research and a case dating back to December 2, 2019 was confirmed by a CT scan and PCR test at the Albert-Schweitzer hospital in Colmar. This work shows that the coronavirus was already circulating in France in early December 2019. It is therefore very likely that other countries were also affected at that time. Thus, the virus was circulating in France before the first cases were officially detected in Wuhan. Which leaves open the question of the origin of this virus: can we really assert that SARS-CoV-2 originated in China?

On January 31, 2020, 220 French returnees from China are placed in quarantine in a holiday camp in Carry-le-Rouet, close to Marseille. At this point, everything still seems under control. But on February 25, a teacher from Crépy-en-Valois, who is also a municipal councillor in Vaumoise, north of Paris, died of a pulmonary embolism at the Pitié-Salpêtrière University Hospital in Paris. The 60-year-old man, tested positive for Covid-19, becomes the first French national to succumb to the disease since the start of the pandemic. The death is a real shock for France: the population realises that the epidemic, although invisible, has started to spread across the country. On February 15, a liturgy celebrated in Mulhouse brings together up to 2 500 people: it will appear to be one of the first significant clusters and one of the starting points of the exponential contamination of the French mainland.

On February 26, Jérôme Salomon, Director-General of Health, speaks to the Senate's Social Affairs Committee: "We have released our strategic stocks and placed an urgent order for protective masks for healthcare professionals. Santé publique France [SPF, the national public health agency—author's note], holds significant strategic stocks of surgical masks. We are not worried about this. So, there is no shortage to be feared, this is not an issue."[25] This assertion is false. The Ministry of Health was alerted in 2018 of the low level of mask stocks but did not react. Actually, France is already short of face masks and protective equipment. On March 17, 2020, the Directorate-General of Health asked SPF to buy urgently 1.1 million of FFP2 masks.

On February 27, Emmanuel Macron visits the Pitié-Salpêtrière hospital. Eric Caumes, Head of the infectious diseases department, warns the President that the situation in France is developing like in Italy because "the virus is

---

[25]Jacquot G (2020, February 27) Coronavirus: « Il n'y pas de sujet de pénurie» de masques, selon le directeur général de la Santé, Public Sénat, https://www.publicsenat.fr/article/politique/coronavirus-il-n-y-pas-de-sujet-de-penurie-de-masques-selon-le-directeur-general.

circulating among us" and "it is probably [sic] transmitted more easily than we thought." Everything was said. What the public still does not know is that France does not have at that time the logistical capacity necessary to follow WHO recommendations and in particular to promote mass testing, due to the limited number of accredited laboratories (only 45 in public facilities) and the limited availability of reagents for PCR testing. This policy will only be corrected on March 28, in order to prepare the lockdown exit strategy.

On March 12, nearly three thousand infection cases are recorded. That evening, during his first television speech devoted to the epidemic, broadcast live by national television channels and radio stations, the Head of State pledges allegiance to science: "A principle guides us in defining our actions, it has guided us from the start to anticipate this crisis and then to manage it for several weeks, and it must continue to do so: it is confidence in science. It's about listening to those who know."

But the government still seems to be hesitant and reluctant, as we have seen, to move on to stage 2 of the epidemic. Information is getting through to the public, but the official announcement is long overdue. Did the economic arguments weigh in the balance? Was it the fear of creating panic among the population? Or the hope for a quick upturn? Most likely. The answer comes four days later, on March 16: in a second televised intervention, the President repeatedly declares that France is in a "health war" against Covid-19 and announces the beginning of a total lockdown period from the 17 March at noon for a minimum duration of fifteen days. "We are at war," the Head of State repeats six times. The French are surprised: what is the connection between the Great War and the coronavirus epidemic? The answer will come several months later: the waves follow, the war goes on, and no one knows when it will end. In any case, this martial tone restores the image of Emmanuel Macron in the public opinion for … a mere couple of days, the time needed for the French population to understand that the country is not ready to face the upcoming health disaster.

The government therefore resigned itself to total lockdown seven and a half weeks after the confirmation of the first patient on its shores, four weeks more than our "Chinese benchmark." This delay seems incomprehensible today, especially as Hubei was already a case study and the number of cases was rapidly growing in France. Greece, for example, closed its schools less than two weeks after recording the first cases and imposed lockdown two weeks later (March 23, 2020). The leaders should have taken into account that, in a geometric (or exponential) sequence, every day counts! See for yourself: on March 2, 2020, France had 191 Covid-19 cases, on March 9, 1 412 and on March 16, the day of the Presidential announcement, 6 633. In one week, the number of cases had multiplied by 7, in two weeks by 35. If lockdown in France had been imposed on March 2, i.e., two weeks earlier, a considerable number of deaths would have probably been avoided. At that time, the government's decision was inevitable but it came too late. A similar hesitation happened a few months later, on July 14, when, noting a resurge of infections, Emmanuel Macron proposed to make mask-wearing in closed places compulsory "in the coming weeks." But that time, the Prime Minister quickly corrected this statement: face masks became compulsory from July 20 onwards, just 6 days after the President's announcement.

The momentum that developed in France at the beginning of March is comparable to the situation described in a publication by a team of scientists led by Nanshan Zhong[26], which should have prompted the government to react more quickly. The Chinese researchers have modelled the development of the epidemic in the three provinces most affected by Covid-19 (Hubei, Guangdong and Zhejiang). Their model reproduces the observed data, and in particular, for Hubei, the epidemic peak of February 20. According to the simulations carried out, a delay of 5 days in the implementation of the containment would have led to a peak around February 25 and the number of contaminations multiplied by four. On the contrary, if the confinement had been brought forward by 5 days, the number would have been approximately halved. These trends and dynamics are no surprise to epidemiologists. Why have these findings not been used more explicitly by the government? By mid-March 2020, three countries, China, Italy and Spain, were already locked down. It must be said that, even in China, the study carried out by Nanshan Zhong's researchers is still not very well known and was not reported in the media. In the UK, according to Professor Neil Ferguson of Imperial College London, whose advice was instrumental when Boris Johnson imposed lockdown on March 23, 2020, the death toll would have been halved had the government decided to go into lockdown one week earlier[27].

In France, the coronavirus crisis develops at a difficult time for hospital staff and emergency physicians in particular. Indeed, they were at the tail end of a six-month strike which started in 2019 and involved almost half of the emergency departments of public hospitals in France. Furthermore, staff are incredulous that, despite repeated assurances to the contrary, there is clearly a shortage of protective masks. Roselyne Bachelot, a Minister of Health from

---

[26]Yang Z et al. (2020, February 28) Modified SEIR and AI prediction of the epidemics trend of Covid-19 in China under public health interventions, Journal of Thoracic Disease, 12(3):165–174, https://jtd.amegroups.com/article/view/36385/html.

[27]BBC News (2020, June 10) Coronavirus: 'Earlier lockdown would have halved death toll', https://www.bbc.com/news/health-52995064.

2007 to 2010, had ordered a stock of 1.7 billion face masks (surgical and FFP2) to fight against the H1N1 influenza epidemic and Jérôme Salomon had assured everyone that a possible shortage was a "non-issue." Yet, where have all these masks gone? After 2010, this huge reserve apparently melted like snow in the sun and was not replaced by Bachelot's successors. "At the time, I was criticised for having bought too many of them, but I assure you that in the event of a pandemic, a billion masks are no luxury," the former Minister explains[28]. Her message is clear: precaution and prevention were not taken seriously by the Presidents who, after 2012, succeeded Nicolas Sarkozy. Prevention of an epidemic does not start when the epidemic starts. Obviously, it is always easy, in the current context, to expose posteriori what should have been done. Two journalists of *Le Monde* were able to trace the famous masks: in May 2017, state stocks stood at 717 million. But in March 2020, there are only 117 million surgical masks left (and zero FFP2-type). Several hundred million masks were burned between 2017 and 2020, overlapping partly with the pandemic. In less than three years, only a sixth of the stock is left[29]. A real "sanitary disarmament," Gérard Davet and Fabrice Lhomme write, in which the government of Edouard Philippe bears a heavy share of responsibility. The reason for this massive destruction? Their expiration date had passed. But, according to the two journalists, this is mainly because at the highest level of State no one has defended the strategic value of the masks (surgical and FFP2). This has been confirmed by the parliamentary commission of inquiry of the National Assembly on the management of the Covid-19 crisis.

Beyond the numbers, the situation is becoming critical and very tense in hospitals and nursing homes. In mid-March, many caregivers make desperate pleas for masks. Their efforts end up paying off: on March 21, 2020, following the request of Jérôme Salomon, the Minister of Solidarity and Health, Olivier Véran, announces an order for 250 million masks—almost two months after the start of the epidemic. However, according to the doctors' estimates, this order will cover the needs for just two weeks… In nursing homes, the situation quickly becomes critical and turns into a hecatomb as the number of deaths has skyrocketed since the end of March. Staff and residents complain about the lack of tests, but it will take until April 6 for Olivier Véran to insist that managers test everyone as soon as the first case of coronavirus appears in their nursing home.

---

[28]Bui D (2020, March 21) Roselyne Bachelot: « Ce que je ressens aujourd'hui ? De la rage !», L'Obs, https://www.nouvelobs.com/politique/20200321.OBS26385/roselyne-bachelot-ce-que-je-res sens-aujourd-hui-de-la-rage.html.

[29]Davet G and Lhomme F (2020, May 7) 2017–2020: comment la France a continué à détruire son stock de masques après le début de l'épidémie, Le Monde, https://www.lemonde.fr/sante/article/ 2020/05/07/la-france-et-les-epidemies-2017-2020-l-heure-des-comptes_6038973_1651302.html.

The speed at which the government reacts, after hesitating for a long time on the strategy to fight the coronavirus, certainly added to these structural difficulties. The political opposition, both left and right, has not shied away from highlighting procrastination over masks, approximations on lockdown, erratic communication and a lack of transparency. It is as if the government had learned nothing from the ongoing epidemic in China, which demonstrated the contagiousness of the virus and the dangerousness of the disease.

More worryingly, the French government is in principle the best informed as it assembled multidisciplinary scientific expertise on no less than four committees. On March 11, 2020, a "Covid-19 scientific council" is set up to guide the government or more exactly, as it states in the official decision, "to inform the public decision in the management of the health situation linked to the coronavirus[30]." Under the presidency of Jean-François Delfraissy, the council initially consists of twelve members (thirteen from April 3, 2020 and seventeen after February 17, 2021). However, one of its members, Didier Raoult, known for his positions on hydroxychloroquine, decides to stop attending its meetings two weeks later, which gives rise to various interpretations. One positive move compared to other countries such as the United Kingdom and the United States is that the opinions of the council are made public[31]. Next, a second committee, called CARE (Comité Analyse Recherche et Expertise, Research and Expertise Analysis Committee), is set up on March 26 to prepare for the phase that will follow the total lockdown[32]. More precisely, the objective of CARE, according to the Ministry of Health, is to "inform the public authorities in a very short time on the follow-up to be given to the innovative proposals for scientific, technological and therapeutic approaches formulated by the French and international scientific community" and "to seek the advice of the scientific community." This second committee, chaired by Professor Françoise Barré-Sinoussi, virologist at the Institut Pasteur and co-laureate of the Nobel Prize in medicine (2008), includes twelve researchers and doctors, two of whom are also members of the Covid-19 council. However, it is not the CARE committee but the Covid-19 council that will speak first on the issue of the lockdown, in an opinion

---

[30] https://solidarites-sante.gouv.fr/actualites/presse/communiques-de-presse/article/olivier-veran-installe-un-conseil-scientifique.

[31] https://solidarites-sante.gouv.fr/actualites/presse/dossiers-de-presse/article/Covid-19-conseil-scientifique-Covid-19.

[32] https://solidarites-sante.gouv.fr/actualites/article/installation-du-comite-analyse-recherche-et-expertise-care.

Fig. 3.2 On March 13, 2020, Prime Minister Edouard Philippe (left), the President of the scientific council Jean-François Delfraissy and the Minister of Health Olivier Véran (right) answer questions from journalists in Paris (© POOL/AFP/Ludovic Marin)

released on April 2, 2020[33]. Moreover, the conclusion of the experts is very vague in this first report: the "decision [to lift the lockdown] could be taken on the basis of epidemiological indicators indicating in particular that the saturation of hospital services, and ICUs in particular, is curbed." A third committee, called "Covid-19 vaccine", is created in early July 2020, under the chairmanship of virologist Marie-Paule Kieny, a former Director-General at WHO and research director at the Institut National de la Santé et de la Recherche Médicale (INSERM). And a fourth committee, set up in 2021, will take care of the vaccination strategy (Fig. 3.2).

Why did the government set up three separate scientific committees to get advice on the management of the health crisis, in addition to the "defence council," which also decides on the actions to be taken in the response to the epidemic? How can the public understand that? And it was yet another committee, HCSP (Haut Conseil de la Santé Publique, High Council of Public Health), which spoke out at the occasion of the hydroxychloroquine controversy. Let us briefly remind ourselves of the facts.

---

[33] Opinion of the Covid-19 scientific council (2020, April 2) Etat des lieux du confinement et critères de sortie, https://solidarites-sante.gouv.fr/IMG/pdf/avis_conseil_scientifique_2_avril_2020.pdf.

A derivative of chloroquine, hydroxychloroquine is an antimalarial medicine which is commercialised and used against lupus or rheumatoid arthritis. It is well known that it causes many side effects (nausea, vomiting, skin rashes, eye damage, heart and neurological disorders, etc.) and can be fatal in cases of overdose. At the beginning of March 2020, several doctors including Didier Raoult, director of the Institut hospitalo-universitaire (IHU) Méditerranée Infection, one of Europe's major competence centres in clinical infectiology, presented hydroxychloroquine as an antiviral remedy against SARS-CoV-2. At that time, this molecule was considered by most healthcare professionals and WHO as a potential treatment but which had yet to be confirmed by scientific protocols. The main argument of specialists against its use at this point is the lack of robust and large-scale clinical trials to determine whether chloroquine and hydroxychloroquine (as well as other molecules) are indeed effective against Covid-19 and whether their use causes side effects.

In the United States, hydroxychloroquine is already being used to treat Covid-19. Trials conducted in China involving around 100 people in different hospitals seem to have shown positive effects. In France, Didier Raoult and his collaborators publish a first study on March 16, 2020, carried out on 36 patients, of which 16, located in Avignon, Nice and Briançon, did not receive any treatment, and 20 in Marseille were treated with Plaquenil (a drug based on hydroxychloroquine sulphate). Marseille scientists observe that after six days, 90% of patients in Avignon, Nice and Briançon are still carriers of SARS-CoV-2, compared to 25% of the patients treated with Plaquenil. They conclude that the treatment, administered at the onset of Covid-19 symptoms, produces a dramatic drop in viral load in four days (compared to twelve). But the article, made public as a preprint on the collaborative platform PubPeer, arouses strong opposition with regards to the applied methodology ("an exceptionally poor experimental design", according to some scientists), in particular because of the very small number of patients, the absence of monitoring for four of those patients, and the fact that the control patients are treated in different clinics. But an American lawyer, Gregory Rigano, succeeded in pitching this "miracle cure" to Fox News, the TV channel President Donald Trump watched every day to get the latest news. This explains how this information went viral…

On March 27, 2020, the IHU team publishes the preprint of a second study, this time on the effects of the combination of hydroxychloroquine and azithromycin, an antibiotic, administered to 80 patients with mild Covid-19 symptoms.[34] Despite the sample size, which is still very small, Didier Raoult

---

[34]Gautret P et al. (2020, March–April) Clinical and microbiological effect of a combination of hydroxychloroquine and azithromycin in 80 Covid-19 patients with at least a six-day follow up: A pilot observational study, Travel Medicine and Infectious Disease, Volume 34, https://www.sciencedirect.com/science/article/pii/S1477893920301319.

concludes that treatment with hydroxychloroquine is significantly associated with a disappearance of the viral load and that its effects are enhanced by azithromycin. Although not a virologist, I am surprised, as a scientist, that this publication was accepted as such, given the absence of a control group and therefore of any point of comparison to estimate the potential impact of the drug. However, reading several articles and the book of Jean-Dominique Michel[35] helped me to understand that, in these times of crisis, some consider it inappropriate to give patients a placebo. As a doctor, Didier Raoult repeatedly said that his top priority is to cure his patients; this is why he refuses to create a control group, whose patients would be deprived of the treatment he considers effective.

After consulting the HCSP, the government adopts derogatory provisions on March 26, 2020, so the day before Didier Raoult's preprint, which authorise the treatment of Covid-19 patients with hydroxychloroquine in line with very strict procedures. Furthermore, the prescription of the antimalarial drug remains reserved for hospital doctors. The HCSP calls upon "prescribing physicians [to] take into account the very limited state of current knowledge and [to be] aware they assume responsibility for prescribing the drugs. The use of hydroxychloroquine is currently excluded for preventive use and treatment." On March 30, 2020, speaking on *France 5* television, the former Minister of Health, Roselyne Bachelot, is exasperated by these scientific quarrels in the midst of a health crisis: "It is a bit the problem of Didier Raoult: he refuses to give his raw data to his peers, he gives the results but not the raw data and that raises a certain number of issues. What I find unfortunate is that Didier Raoult is a great infectious disease specialist, and a recognised scientist in this field. However, actually, he is his own worst enemy, and the worst enemy of his scientific method. I think if he had a little friendliness, respect for other professionals who are eminent personalities too, it would work better." But the minister, herself a scientist, admits with some candour that which might be interpreted as bad faith: "What I find most regrettable is that behind this debate there is suffering, there are people waiting for treatment, there are people who see their loved ones dying, and who see this debate which looks like a ragpickers' quarrel".[36]

---

[35]Michel JD (2020) *COVID: anatomie d'une crise sanitaire*, humenSciences, Paris.

[36]Rouyer M (2020, March 31) "Je vais cogner": Roselyne Bachelot agacée par les querelles scientifiques en pleine crise sanitaire, Gala, https://www.gala.fr/l_actu/news_de_stars/video-je-vais-cogner-roselyne-bachelot-agacee-par-les-querelles-scientifiques-en-pleine-crise-sanitaire_445805?utm_term=Autofeed&utm_campaign=FbGala&utm_medium=social&utm_source=Facebook&fbclid=IwAR2wyrW4HxHxReH4AHe_4VBHr3gXYTflOxKMGBD51VSbMPZh8LVPftywQRs#Echobox=1585639362.

In the public domain, the scientific study of hydroxychloroquine has turned into a political controversy. While no one disputes the need to follow scientific protocols in order to establish that the health benefits of a new drug outweigh its known undesirable effects and unknown risks, the exchanges between experts looked rather academic and oblivious to the issue at hand. However, a doctor cannot argue that the health emergency justifies the use of a drug for which we do not have all the necessary guarantees of effectiveness although its safety has been verified for a long time. Furthermore, GPs do not understand why it is impossible for them to prescribe hydroxychloroquine and patients do not understand it either. But in practice, doctors cannot guarantee the beneficial effect of the molecule. A step forward, a step back: the government's approach is not crystal clear and is strangely mixing-up genres—political and scientific/medical—notably with the visit of Emmanuel Macron to the IHU in Marseille on April 9, 2020. Everyone benefits when there is a clear and tight separation between the rational evaluation of therapeutic substances and the sometimes-illogical urgency of political decisions. But this controversy has clearly shown how anxious top-level decision-makers can be: under intense pressure, they have only one priority in mind: to be able to report on any progress or announce a possible treatment in the fight against the disease. Didier Raoult, despite his many contributions and his undisputed professional reputation, cannot place himself above the law of science.

Against this context, the public is not ready to hear that the approval procedures for new drugs or new vaccines are quite long: once they are synthesised, the molecules must first be tested in vitro and then tested on living organisms, probably on animals and finally on human volunteers before undergoing, if the results are conclusive, large scale trials and then, hopefully, commercialisation. During a crisis, time always tends to accelerate. Because of the pandemic, many people are calling for scientific procedures and publications to cut corners. A real "scientific populism" tends to emerge: rationality is seen as a constraint and sparks criticism about clinical trials, standardisation, randomisation, and data transparency. Still, most scientists and doctors are not prepared to take risks for human health—even in an emergency situation.

Another controversy then develops in France at the end of March 2020: face masks. Since the start of the epidemic, WHO and many governments including France repeat that masks should only be used by caregivers, patients and their close contacts, according to the available scientific information.[37] The then government spokeswoman, Sibeth Ndiaye, said over and over again

---

[37] However, early June 2020, WHO adapts its recommendations and advocates "non-medical, fabric masks for use by the general public when physical distancing cannot be maintained, as part of a comprehensive 'Do it all!' approach, including improving ventilation; cleaning hands; covering

that wearing a mask is not necessary and may even be counterproductive. But on March 31, 2020, *Le Monde* publishes the translation of an interview given by George Gao, Director-General of the Chinese CDC, to the American journalist Jon Cohen for *Science*. The scientist, whose team isolated and sequenced SARS-CoV-2, says in the interview that "The big mistake in the U.S. and Europe, in my opinion, is that people aren't wearing masks. This virus is transmitted by droplets and close contact. Droplets play a very important role—you've got to wear a mask, because when you speak, there are always droplets coming out of your mouth. Many people have asymptomatic or presymptomatic infections. If they are wearing face masks, it can prevent droplets that carry the virus from escaping and infecting others."[38]

As of April 3, 2020, the American health authorities are advising the population to go out with their faces covered. France is also changing its position, as Jérôme Salomon is now encouraging French people to wear "alternative" masks (but which were not yet available at that time!). The government has since done an about-turn and now believes that masks can be worn. Several cities, such as Nice, even started to impose mask-wearing outside. On July 20, wearing face masks becomes compulsory indoors throughout France. The mess is actually camouflaging a major problem: mask shortages. Again, this turns out to be incomprehensible. In 2005, the Parliamentary Office for Evaluating Scientific and Technological Choices (OPECST) adopted a report which already concluded that wearing a mask is a very effective instrument to fight against a large-scale epidemic, as well as reassuring the population.[39] Fighting against a pandemic, the report recommends, among other measures, setting up physical barriers: "This method, which is very effective, involves the possibility of closing schools, prohibiting gatherings and limiting public transport in large cities; it also implies that people in contact with the public can wear masks adapted to the pandemic." The report covers all the main points that we are hearing about today. More specifically, it states that "the provision of masks in sufficient numbers would certainly have a very high cost but, at the same time, would help limit the paralysis of the country. Seen

---

sneezes and coughs, and more." WHO recommends examining the precise conditions on a case-by-case basis, which depend on the circulation of the coronavirus, the professional categories concerned, etc.: https://www.who.int/emergencies/diseases/novel-coronavirus-2019/question-and-answers-hub/q-a-detail/q-a-on-Covid-19-and-masks.

[38]Cohen J (2020, March 27) Not wearing masks to protect against coronavirus is a 'big mistake,' top Chinese scientist says, Science, https://www.sciencemag.org/news/2020/03/not-wearing-masks-protect-against-coronavirus-big-mistake-top-chinese-scientist-says.

[39]Door JP and Blandin MC (2005, May 10) The epidemic risk, rapport n°332, Sénat, https://www.senat.fr/notice-rapport/2004/r04-332-1-notice.html.

from this angle, the cost should be put into perspective." A very explicit plea for "strategic" masks.

These communication flip-flops give an impression of indecision and a day-to-day lack of preparation. This is, after all, the main characteristic of a crisis: what is decided today can be put into question tomorrow and no one knows what will be implemented the day after tomorrow. Was the government really listening to doctors and scientists? Despite a profusion of committees, it is not certain that the weight of scientific expertise affected the balance of executive decisions in the end. Let's not be naive: these expert groups also act as a "fuse." When the tension is mounting, the government can call on scientists to help restore a sense of trust by sweetening the pill with their "scientific-sounding" pronouncements and their positive public image. The Minister of Health had also made it clear that the opinion of the Covid-19 council is not a recommendation, just one opinion among many.[40] At best, the council and the various committees are a source of ideas, at worst a facade or even a screen or a shield. This is clearly noticeable on March 18, 2020, when Jean-François Delfraissy speaks at length on *France 2*'s main evening news and takes it upon himself the criticisms addressed to the government for its lack of foresight. "Why, given the situation in China and Italy, did we wait a fortnight too long before going into lockdown?", asks Anne-Sophie Lapix. "Perhaps we did not sufficiently appreciate the gravity of the epidemic," Delfraissy courageously concedes. This "shielding effect" is frequently observed during health crises. In the midst of France's infected blood scandal (1991) for example, Sophie Coignard and Alexandre Wickham[41] quote extracts from a note of an adviser addressed to the director of the cabinet of Minister Edmond Hervé: "I think it is better for the moment that it is the luminaries who speak on this subject. The minister could then say: "The experts think that it was impossible at the time to avoid these contaminations"." In our society, the myth persists that science is pure and objective (with the—at least passive—support of scientists who benefit from a "perfect impunity bubble"). However, this perception can only lead to misunderstandings and errors.

There is no doubt that managing a major crisis must be one of the worst jobs for politicians and communicators. You have to manage the situation in the face of uncertainty, work with a sense of urgency with all the agencies concerned, make critical decisions in the clearest and most consensual way

---

[40]Ribière C (2020, March 25) Coronavirus: face au Conseil scientifique, Macron a le dernier mot, Le journal RTL, https://www.rtl.fr/actu/politique/les-infos-de-12h30-coronavirus-face-au-conseil-scientifique-macron-a-le-dernier-mot-7800310713.

[41]Coignard S and Wickham A (1999) L'Omerta française, p. 271, Albin Michel, Paris.

possible and communicate with maximum transparency all speaking with one voice. A huge challenge! Under these conditions, the decision to put the experts of the Covid-19 scientific council and the CARE committee on a stage to speak in front of the public and the press, was a double-edged sword. There is a risk that the words of the scientists will be taken literally and their subjective opinions believed to be facts. However, in a crisis, scientists do not necessarily have the *right* science. Obviously, putting the experts in front of the cameras reinforces transparency and shows that the government takes their opinions into account. But this type of communication can also be interpreted as a demonstration that central government is indecisive and uncertain, hides behind scientists and does not take responsibility for its actions. This is what happened with the mask controversy. From March to July 2020, the President, the Prime Minister and the government spokesperson were often talking to the media without consulting each other. While more and more people wear a mask in the street, the spokesperson says on March 20, 2020: "I don't know how to use a mask. I could say, 'I'm a minister, I put on a mask,' but actually I don't know how to use it! Because the use of a mask involves precise technical gestures." Communication is only the tip of the iceberg: if your messages are inconsistent, so is your strategy.

Agnès Buzyn, an oncologist by training and appointed Minister of Solidarity and Health on May 17, 2017, was working on the front line of the government when the coronavirus exploded in France. After her resignation from the government on February 16, 2020 to run for mayor of Paris, she publicly underlines the hiccoughs in crisis management and government communication. In an interview given on March 17 to *Le Monde*, she throws a stone in the political pond by admitting to having alerted the President of the Republic and his Prime Minister in January of the seriousness of the Covid-19 epidemic. Her confession is terrible: "On January 11 [when China announced its first death—author's note], I sent a message to the President about the situation. On January 30, I warned Edouard Philippe that the [municipal] elections could probably not be held." In short, the former Minister argues that the French political elite did not listen to her warnings. Maintaining the first round of municipal elections on March 15, as the epidemic spread through France, was a "farce," she says. And she accuses France of not having followed the precautionary principle, yet enshrined in the Constitution, which applies in cases of uncertain risk (i.e., for which the calculation of probabilities is impossible). But this contradicts the fact that on January 24, 2020, during a press briefing, the minister was clearly reassuring: "The risks that the virus is spreading in the population are very low," she declared. How could she have made such a statement knowing what was happening in Wuhan? Was the government sure that it was in control? Or did the minister want, at all costs, to assure (or reassure) the people of France? Either possibilities should worry us.

Faced with a general outcry provoked by the interview, Agnès Buzyn corrected some of her declarations in a formal press release,[42] without contradicting her main point: "Agnès Buzyn regrets the tone of this article [of *Le Monde*] and the use that has been made of it at a moment where the whole country should concentrate on crisis management. She considers that the government has been fully up to the challenge to face this virus." She also adds that in January, there was no serious alert from WHO and no consensus among doctors on the pandemic spread of the virus. Which is not true, as we have seen.[43]

An article in *Liberation* gives a very different version of events. According to the journalist who wrote it, Charlotte Belaïch, the Ministry of Health was informed of the situation from the very first hours of the epidemic. At the end of December 2019, the ministry's health watch and safety unit had already identified several suspected cases of acute respiratory infections in Wuhan. On January 2, 2020, the Operational Centre for Reception and Regulation of Health and Social Emergencies (Corruss) in the Ministry of Health began to watch the situation more closely. At this time, scientists are still wondering if transmission between humans is possible, but the confirmation that it is comes on January 20. On January 29, an interim memo states that there is a "moderate risk" of importing cases into Europe and a "very high" risk of secondary cases in the event of late detection or lack of prevention and management measures.[44] It is the next day that WHO launched the maximum alert. But the organisation also states that "The [emergency] committee believes that it is still possible to interrupt virus spread, provided that countries put in place strong measures to detect disease early, isolate and treat cases, trace contacts, and promote social distancing measures commensurate with the risk." Agnès Buzyn reportedly informed the Prime Minister the same day.

---

[42] https://www.parismatch.com/Actu/Politique/Coronavirus-et-municipales-Agnes-Buzyn-retropedale-et-precise-ses-propos-1679164.

[43] Let us recall here that WHO convened a meeting on January 22, 2020 to determine whether or not the outbreak constitutes a PHEIC (public health emergency of international concern). The experts had mixed feelings at that time. But one week later, WHO decided to qualify Covid-19 as a PHEIC.

[44] Belaïch C (2020, April 4) Covid-19: Agnès Buzyn, les failles d'une ministre et d'un système, Libération, https://www.liberation.fr/politiques/2020/04/04/Covid-19-agnes-buzyn-les-failles-d-une-ministre-et-d-un-systeme_1784003.

Thus, as of January 29, 2020, the government had in principle all the cards in its hand. Admittedly, there were still uncertainties about the risk of a pandemic and the situation was evolving rapidly, but the epidemic in China, the information from WHO and the warnings from the scientific community were clear enough to predict what was to come. France will not impose full lockdown until almost two months later. "The government should have been warned earlier," said Socialist MP Joël Aviragnet at the end of February, deploring the lack of information given by the executive.

A senior official told *Liberation* his surprise that the "pandemic plan," established in 2009 at the time of the H1N1 influenza, had not been activated by the government. This plan is a decision aid that is triggered in the event of serious risk or public health crisis. In the case of Covid-19, an interministerial crisis unit (CIC) was indeed set up, but only on March 17, the first day of lockdown. "An aberration," according to the official. The government is apparently responsible for this late decision, as they did not want to create any panic in the French population. Remember what the Wuhan Health Commission told Dr Ai at the outset of the epidemic: "If there is panic, you will be responsible."

Regarding face masks, it seems that Agnes Buzyn only ordered 50 million when she was in office, while the needs are close to 40 million per week at level 3 of the epidemic. The National Union of Nursing Professionals (SNPI) speaks of a "state scandal."

Was Agnès Buzyn speaking out of both sides of her mouth, one side for the public and the other for the government? Anticipating possible legal investigations, was she trying to present the work she did in the ministry in a favourable light? When she left the ministry on February 16, did her tears reflect a feeling of failure? I contacted Charlotte Belaïch, the journalist of *Liberation* by email on April 8, 2020 who replied: "My feeling is that the Prime Minister and the President did not take the measure of the crisis to come in the months of January and February and that Agnès Buzyn, for her part, was not armed to manage the crisis without the support of the Head of State…".

On June 2, 2020, a special commission of the National Assembly, one of two chambers of the French Parliament, starts investigating the government's response to the Covid-19 pandemic. Initially, the role of the commission is restricted to monitoring the government's response to the pandemic during the state of health emergency that was declared on March 23, 2020. However, on June 2, its prerogatives are expanded to become a full investigatory commission and study all aspects of the "impact, management, and consequences of the coronavirus epidemic." During her hearing on June 30, 2020,

Agnès Buzyn did not say anything new.[45] She confirms that she "anticipated" the health crisis and noted in January the insufficiency of the number of masks available, adding that "the stocks are not managed at the level of the minister." France was "extremely well prepared" for the epidemic, she says. In a shaky voice, the former minister hesitates, searches for her words and, in fact, does not succeed in convincing the members of the Parliament. On masks, the President of the commission, Brigitte Bourguignon, evokes a letter sent on September 26, 2018 by François Bourdillon, then CEO of SPF, to Jérôme Salomon in which he proposes to reconstitute a stock of a billion masks [to provide twenty million French households with fifty masks—author's note] and underlines that 95% of the antiviral drugs stored by the national agency are out of date. Apparently, there has been no follow-up. It seems that a first order of 1.1 million FFP2 masks was placed on January 30, 2020. But SPF got the information on March 17, 2020, almost two months later. Health officials, including ministers, did not act in an anticipatory manner. At the start of the crisis the stocks of masks had been reduced from 2.2 billion in 2009 to 117 million. The rapporteur of the commission, Eric Ciotti, admits having been intrigued by a handwritten note clearly visible in the file that Agnes Buzyn had put in front of her during her hearing, on which it was written: "Say as little as possible about the masks."[46]

On October 21, 2020, it is the turn of former Prime Minister Edouard Philippe to be cross-examined by the commission. While conceding a few errors, he implies that he was poorly advised. He also admits that France was insufficiently prepared for the arrival of an unknown virus. "I made communication mistakes, like many others, as we can all do," he declares. However, the former prime minister attributes these errors to the medical doctrine that prevailed at the time. "I said that wearing a mask made no sense for the general population. Why did I say it? Because I had been told. Because the doctors told me. Because that was the doctrine, which didn't change until much later. And I said it confidently because I thought it was necessary to disseminate this information which resulted from the doctrine. Today the doctrine has changed. Each time people question me about this episode, I answer by a question: "Should I have said it a lot more carefully to protect myself? Or should I have to say it in the most convincing way because that

---

[45] Full transcript of the hearing on June 30, 2020 of Mrs Agnès Buzyn, former Minister of Solidarity and Health, before the national Assembly's information mission on impact, management and consequences of coronavirus Covid-19 epidemic in all its aspects: https://www.youtube.com/watch?v=sYQmSMTSdis.

[46] Hecketsweiler C and de Royer S (2020, July 28) « Notre système de santé a bel et bien été débordé» par le coronavirus, Le Monde, https://www.lemonde.fr/politique/article/2020/07/29/eric-cio tti-notre-systeme-de-sante-a-bel-et-bien-ete-deborde_6047603_823448.html..

was the doctrine?" I still do not have the answer to the question. I tend to think that when you are managing a crisis, it is better to think about the decision you take and the message you convey rather than what will happen to you three months later. So, I take responsibility for what I said, although I see clearly that I appear to be in an awkward situation." Edouard Philippe finally regrets that the public debate around these medical and scientific questions has not been able to take place peacefully. According to him, this comes from the fact that the government was not backed up by solid and respected expertise [sic] on this matter in order to legitimise its decisions. "I am convinced that constant invective and violent criticism on complex and uncertain issues, has made our fellow citizens reluctant to contribute to the fight against the coronavirus," he notes. And concludes: "I don't know how to correct that."

There is therefore a set of coherent arguments that support the idea that the government was well informed of the possible risks and could—and should—have taken precautionary and preventive measures much earlier. If lockdown had been instigated when the first red flags appeared at the end of January 2019, the number of cases and deaths could have been divided by a thousand. That is to say around 100 deaths… Of course, it is easy to put forward such a figure today. But let us not forget either that at the end of January, Hubei was already in lockdown and various studies and simulations were circulating in the scientific community and in government circles.

Has the lifting of lockdown been managed any better? On April 13, 2020, Emmanuel Macron causes a stir by announcing the "progressive" exit from national lockdown on May 11, 2020, without clearly setting any conditions. Above all, this exit appeared to be extremely fast. In Wuhan, the lockdown (which was much stricter than in Europe) was two and a half months long (from January 23 to April 8). In France, it will be nearly three weeks shorter. In its opinion issued on April 2,[47] the Covid-19 council establishes a long list of measures to be taken, such as "the availability of material protections" and "methods to isolate cases and their contacts adapted to the personal context." It seems clear, at that time, that the relaxation of restrictions is almost impossible before the first half of May, at best. But the government announces the date of May 11, against the advice of the scientific council.

Then, another communication hiccough follows: on April 15, Jean-François Delfraissy indicates that people "over 65 or 70" should remain in isolation. The President of the council is then immediately corrected by Emmanuel Macron who declares on April 17 that he does not want to

---

[47] Opinion of the Covid-19 scientific council (2020, April 2) Etat des lieux du confinement et critères de sortie, https://solidarites-sante.gouv.fr/IMG/pdf/avis_conseil_scientifique_Covid-19_8_avril_2020.pdf.

discriminate between citizens after May 11 and calls for individual responsibility. Even the man on the street can see that statements on such an
important subject should have been coordinated and subject to internal
discussions before being brought into the public arena.[48] On April 19,
Edouard Philippe announces during a press conference lasting nearly three
hours, that the "principles" of the lockdown phase-out which include the
increased production of masks, a target of half a million tests per week and
the possible mandatory quarantine of patients who test positive for the illness
(in particular in hotels). At last, France seems to have learned the lessons from
South Korea, Germany and China.

On April 20, the scientific council publishes its opinion on the lockdown
phase-out.[49] The experts propose in particular "to keep nurseries, schools,
colleges, high schools and universities closed until September." The council
justifies its opinion by the fact that "the risk of transmission is significant
in places of mass gatherings such as schools and universities, as it is particularly difficult to adopt barrier measures among the youngest. However, a
"note" dated four days later corrects the council's proposal: "The scientific
council took note of the political decision to carefully and gradually reopen
schools from May 11, taking into account health issues as well as societal and
economic issues."[50] Had the government asked the council to re-examine
its opinion? Or had the President decided to keep a distance from science
on this point? In any case, it shows that the council modified its opinion
in the light of the "political decision." This is not what we expect from a
scientific advisory board. It is therefore difficult to believe in the sincerity
of Olivier Véran when he makes assurance, in an interview on May 2, that
the government and the scientific council are "in agreement."[51] The council
attaches a long list of conditions to its conclusion for the lockdown's exit,
which demonstrates that it is a high-risk operation. It should be noted in
passing that the strategy recommended by the experts is along the same lines

---

[48]This seems to echo the results of an opinion poll carried out in Italy during the epidemic: 48%
of the people interviewed think that the diversity of advice publicly given by experts has created
confusion: https://sagepus.blogspot.com/2020/04/italian-citizens-and-Covid-19-one-month.html.

[49]Opinion of the Covid-19 scientific council (2020, April 20) Sortie progressive de confinement
– Prérequis et mesures phares, https://solidarites-sante.gouv.fr/IMG/pdf/avis_conseil_scientifique_20_
avril_2020.pdf.

[50]Note of the Covid-19 scientific council (2020, April 24) Enfants, écoles et environnement familial
dans le contexte de la crise Covid-19, https://solidarites-sante.gouv.fr/IMG/pdf/note_enfants_ecoles_
environnements_familiaux_24_avri_2020.pdf.

[51]Beaumont O, Méréo F and Mari E (2020, May 2) « La date de levée du confinement pourrait
être remise en question», Le Parisien, http://www.leparisien.fr/politique/olivier-veran-la-date-de-lever-
du-confinement-pourrait-etre-remise-en-question-02-05-2020-8309694.php.

as that which was followed by the Chinese authorities in Wuhan: identification of cases at the earliest possible stage with quarantine followed by contact tracing, and systematic testing with possible isolation of these contacts even if they are asymptomatic. Cynicism or realism: Marine Le Pen, President of the National Rally (Rassemblement National, far-right political party), has this harsh conclusion: "The scientific council is a bit of a guarantee: we take it out when it suits us, and then we put it under the carpet when it does not suit us."[52]

However, the government troubles are not over yet: on April 30, Olivier Véran presents the lockdown exit map for the first time, with all departments coloured in red, orange or green depending on the two indicators used to assess the risk of a second wave: the circulation of the virus (in reality the percentage of cases in the population) and the hospital pressure in intensive care (in reality the percentage of Covid-19 patients in ICUs). Unfortunately, the map contains several errors, and people will see nine departments changing colour overnight... There is also a lack of *transparency*: the coding of the indicators is not disclosed. And a last hiccough: the Secretary of State for Digital, Cédric O, announces on May 5 that the release of StopCOVID, the French smartphone application intended for tracking Covid-19 patients, is postponed until June 2. Two months later, just 2.5 million people have downloaded the application and less than 3% of the population claim to use it. It's little consolation but contact tracing apps are unsuccessful in many countries.

It is therefore in an atmosphere of confusion and anxiety that France cautiously eases its lockdown. The French know this is going to be a high-risk operation, but they are concerned by the uncertainties that the government is not addressing: are the masks going to be available in sufficient numbers? Is the target of 500 000 weekly tests achievable? How do you explain the fact that in departments labelled "green", four new infection "clusters" appeared the day before the end of lockdown? Last but not least, the model predictions of the evolution of the epidemic after May 11 are bleak and indicate that a second wave of the epidemic seems inevitable..[53] And to close the loop, health professionals are raising concerns and questions about the overall

[52]Lemarié A and Faye O (2020, May 14) Entre le conseil scientifique et l'exécutif, une relation aigre-douce, Le Monde, https://www.lemonde.fr/politique/article/2020/05/14/entre-le-conseil-scientifique-et-l-executif-une-relation-aigre-douce_6039603_823448.html.

[53]Benkimoun P and Hecketsweiler C (2020, May 7) A la veille du déconfinement, des projections épidémiologiques globalement pessimistes, Le Monde, https://www.lemonde.fr/planete/article/2020/05/07/a-la-veille-du-deconfinement-des-projections-epidemiologiques-globalement-pessimistes_6038921_3244.html.

public health strategy. Where is the balance between individual responsibility and imposed restrictions? How can infections be stopped as quickly as possible in places of high density such as families, residences and some workplaces? While scientific knowledge is still fragmented, it is nevertheless progressing rapidly and the impression is emerging that the authorities do not sufficiently take into account the lessons learned in other countries. Still, the return to "another normal life" begins on May 11, 2020, with progressive lifting of restrictions, relaunch of economic activity, with the exception of bars and restaurants, and the gradual opening of classes.

On June 5, Jean-François Delfraissy declares on *France Inter* radio that the epidemic is "under control." His words echo the note released by the scientific council two days earlier, which establishes four likely scenarios for the rest of the epidemic taking into account the current situation and the knowledge gained. The first and most favourable of these scenarios indeed predicts an "epidemic under control." The second one anticipates the appearance of "critical clusters;" the third refers to a "gradual and quiet resumption of the epidemic;" and the fourth describes a critical degradation of the indicators which would result in "a loss of control of the epidemic."[54]

A month later, on July 9, the infectious diseases' specialist gives a somewhat different speech. Jean-François Delfraissy expects a new wave of the epidemic in Europe due to the too rapid lifting of preventive measures. He still supports the lockdown approach despite the many criticisms and its multiple side effects at the level of both individuals' health (postponed medical treatments, stress, etc.) and social conditions (unemployment, setbacks in education, etc.): "I remain convinced that it was not the right solution, but the least worst."[55] So, why did the authorities let their guard down over the summer? Why didn't they encourage the population to be more cautious by sending more precautionary alerts? Why did they stick to the line that testing and isolation would stop the epidemic? This kind of light-touch approach probably reflects both a lack of professionalism on the side of the government and the experts and the collective error of the population given the silent majority embraced this way forward.

And the resumption of the epidemic approaches. On July 14, 2020, Emmanuel Macron announces that wearing a face mask in enclosed public places will be compulsory from August 1. I immediately respond to his tweet:

---

[54]Opinion nr 7 of the Covid-19 scientific council (2020, June 2) 4 scénarios pour la période post-confinement – Anticiper pour mieux protéger, https://solidarites-sante.gouv.fr/IMG/pdf/avis_conseil_scientifique_2_juin_2020.pdf.

[55]Hecketsweiler C (2020, July 9) « Nous sommes à la merci d'une reprise de l'épidémie en France», Le Monde, https://www.lemonde.fr/planete/article/2020/07/09/jean-francois-delfraissy-nous-sommes-a-la-merci-d-une-reprise-de-l-epidemie-en-france_6045651_3244.html.

"But why not immediately?" On July 15, the map published by SPF shows that the R_t factor is close to or greater than 1 in all regions but I do not see any particular reaction in the media. Mind blowing: again, everyone seems to have forgotten the exponential law… And where is the scientific council? Only this statement from the Prime Minister: "If we notice before that date [August 1—author's note] that the epidemic is resuming, we would bring the deadline forward, but it is not worth worrying the population," Jean Castex (who succeeded Edouard Philippe as Prime Minister on July 3, 2020) says the same day. The next day, July 16, he makes a correction: "Wearing a mask in enclosed public places will be compulsory from *next week*." A positive point: the increase in the number of cases is reflected in a relatively moderate resumption of deaths, which emergency physicians explain by earlier treatment of severe cases and a slight improvement in the treatments applied. On July 21, Jean-François Delfraissy speaks for the first time about the new wave: "The figures are not good, they are worrying." He is talking about a "slow resumption" which the health authorities attribute to "a lifting of the barrier gestures […] and a lower adherence to physical distancing in particular." A very neutral tone which could possibly be interpreted as a denial of the government's analysis of the health crisis. On August 16, with more than 3 000 new infections in 24 h, the Minister of Labour, Employment and Economic Inclusion announces that "before the end of August, new health measures will apply in companies." Again, why such late implementation? Since August 5, the number of cases is practically doubling every week. But it is not until August 28 that the Directorate-General of Health recognises, in its daily press release, that "the dynamics of the epidemic's progression is exponential" across France. After Toulouse and Marseille, the generalisation of mask-wearing spreads to Paris, Strasbourg and Bordeaux. Should people prepare for local or regional containment plans? Still no information from the scientific council. In the media, experts rightly point out that the indicators, especially the absolute death toll, are well below those for March–April. Is this normal or an exponential trajectory? What should the public conclude? One should not forget here that a trend is not measured only by the absolute numbers but by its evolution (variation, slope, etc.). On September 6, SPF confirms that "in mainland France, the progression of viral circulation is exponential. The dynamic growth of transmission is worrying." But apart from mandatory mask-wearing in the big cities, no information comes out from the government's scientific committees. Unfortunately, this pandemic is a race against time: every week lost could be catastrophic.

Finally, the scientific council breaks its silence on September 9, with a new opinion (written on September 3) devoted to the methods of isolation. The council points to the lack of respect for the quarantine fortnight by infected people. Remember that in France isolation is only "recommended" in the event of a positive test. Noting "the non-compliance with isolation measures by a potentially large part (...) of people infected or by contacts," the experts recommend shortening the period of isolation to seven days, rather than fourteen. The council says this is based on scientific data showing that most cases are contagious within ten days: four days before the symptoms appear and six days after.[56] One cannot exclude the fact that the council also took into account with perhaps a fair amount of fore-thought the impact on the economy as an increasing number of companies across France are facing the risk of bankruptcy. The council takes note that in Europe, contrary to Asia for example, you cannot manage an epidemic by relying on the people's goodwill nor can you impose harsh constraints. Therefore, the scientific council does not recommend more restriction measures but instead proposes a subtle combination of "the obligation of solidarity (through self-isolation) with strong compensation measures." However, Jean-François Delfraissy announces in a somewhat cryptic way that "the government will have to make difficult decisions" within ten days.

The next day, on *RTL radio*, the President of the council refuses to say more about these decisions except that he is opposing the idea of a new lockdown. I send him a tweet: "Dear Professor, how can you, as a representative of a public body, communicate in such a non-transparent way?" Two days later, Prime Minister Jean Castex confirms that the fortnight will be reduced to a "septaine" (a seven-day quarantine). But he refuses to take more coercive measures suggested by the President of the scientific council. An official close to the government explains: "The scientific council gives scientific advice, the executive must make the country work by taking into account all the parameters, especially the social ones." Throughout this period, government communication focuses on the number of cases and tests—for which requests are piling up. However, there is no information provided about the fact that the data, at that time, indicates a decrease in the case fatality rate. Despite the increase of the number of Covid-19 cases, the death toll remains relatively low, and not only in France. Is it due to a mutation in the virus? Or the fact that a growing part of the population has protective immunity? Is it because physicians are getting used to treating the disease and are providing

---

[56]Opinion nr 9 of the Covid-19 scientific council (2020, September 3) Stratégie et modalité d'isolement, https://solidarites-sante.gouv.fr/IMG/pdf/avis_conseil_scientifique_3_sept embre_2020.pdf.

better patient care? Or is it the just the consequence of generalised preventive and barrier measures that keep the number of severe cases relatively low?

That same day, I join a group of researchers and doctors led by Jean-François Toussaint, professor of physiology at the University of Paris, and Laurent Mucchielli, a sociologist at CNRS, to denounce the government's communication in a column published by *Le Parisien*.[57] We want to pass on the message that government communication is mainly about "systematically [exaggerating] the dangers without explaining the causes and mechanisms." Bottom line, we call for re-scoping or replacing the scientific council because currently it does not guarantee "transparency, pluralism, open debate and evidence-based decisions." The official communication, which we qualify as "anxiety-provoking" and "counter-productive" and, let's say, unprofessional, goes precisely against the objective that the government has set for itself: "To make people responsible without making them guilty or childish." We all believe that, at a time when the majority of citizens no longer trust official discourse and when conspiracy theories of all kinds abound on social networks, government communication on Covid-19 must be complete and transparent. This is the message I always advocated for during my career as a science communicator, in particular when I was the head of communication at ITER.[58] Don't underestimate the public's capacity to dig into scientific and technical details! Don't hide any negative points, as they will find them! Be as open as possible about the benefits and risks (real and perceived) of the strategy or the policy that you propose.

The aim of our initiative was to start a public debate and answer very basic—but scientific—questions such as: what are we measuring exactly? What do the published figures used by the government to legitimise its action really mean? In this respect, the group advises against focusing only on the number of positive tests since physicians do not have always detailed information on the severity of these cases. Trends in the number of deaths, hospital admissions and ICU admissions are therefore more likely to indicate whether or not the epidemic has resumed. But what is more serious is that the prefectural decisions on restrictive measures are based on the incidence rates, i.e., the number of positive cases per 100 000 inhabitants. However, these rates are not normalised by the number of tests. A mistake most high school students

---

[57]Toussaint JF et al. (2020, September 10) Covid-19: nous ne voulons plus être gouvernés par la peur, Le Parisien, https://www.leparisien.fr/societe/Covid-19-nous-ne-voulons-plus-etre-gouvernes-par-la-peur-la-tribune-de-chercheurs-et-de-medecins-10-09-2020-8382387.php.

[58]ITER (International Thermonuclear Experimental Reactor) is a collaborative project funded by China, the EU, India, Japan, Russia, South Korea and the United States. The seven partners are jointly building an experimental fusion machine (tokamak) situated in the south of France. The ultimate aim is to demonstrate that nuclear fusion could become a new source of energy on Earth.

would not make because if you increase the screening capacity, your number of positive tests will automatically increase: a sure-fire way to fabricate a second wave from scratch!

Another technical issue: what do PCR tests measure exactly? In France, these molecular diagnostic tests do not mention the number of cycles (CT, cycle threshold) used, nor the virus concentrations. Rather, the PCR results depend on the number of DNA amplification cycles required to observe them by fluorescence. If this number is too low, the tests will not detect the virus; if it is too high, the presence of a few viruses will be enough to make the test positive while the patient might not be contagious. With a CT between 40 and 45, as is the case in France, or between 37 and 40 as in the United States, PCR tests detect virus fragments or a very small amount of virus. It cannot be said that these are "false positives" since viruses or fragments are indeed present, but probably in insufficient quantity to consider that the patient is contagious. With sophisticated and powerful tools, you will find anything! According to a *New York Times* article published on August 29, 2020, nearly 90% of positive tests would not have been so with a cycle count set at 30.[59] "Any test with a cycle threshold above 35 is too sensitive," some specialists say. But the most astonishing thing is that the CT varies from one device to another and from one laboratory to another. And data on the number of cycles of PCR tests does not appear in the statistics published by the French health authorities (nor in the United States for that matter). This casts some doubts on government policy, based on the number of new cases. Surprisingly, these technical discussions take place more than six months after the start of the epidemic. And why is there no international harmonisation?

At the same time, *Le Monde* publishes a survey carried out among some fifty doctors, infectious diseases' specialists, virologists and epidemiologists.[60] Their responses demonstrate the scientific consensus that has formed around Covid-19: they all believe that significant progress has been made in understanding the disease, the spread of which can be contained with the help of simple and inexpensive measures. On the other hand, the researchers are cautious about the effectiveness of wearing masks outdoors, but recognise that it is a justifiable precautionary measure. Even if a majority of the scientists interviewed believe that the actions implemented are not disproportionate,

[59]Mandavilli A (2020, August 29) Your Coronavirus Test Is Positive. Maybe It Shouldn't Be, The New York Times, https://www.nytimes.com/2020/08/29/health/coronavirus-testing.html?auth=login-email&login=email.

[60]Audureau V, Dagorn G, Maad A (2020, September 30) Covid-19: comment les scientifiques jugent la stratégie sanitaire française, Le Monde, https://www.lemonde.fr/les-decodeurs/article/2020/09/30/masques-mesures-communication-comment-les-scientifiques-jugent-la-strategie-sanitaire-face-au-Covid-19_6054250_4355770.html.

they are on the other hand very critical of the massive screening strategy with its excessive delays. They also deplore the way the government is communicating, believing it to be not up to the task and responsible for fuelling fear.

We don't realise it until later but, in fact, at the end of September, the scientific council had suggested to the government to put in place a curfew for about two weeks in the main cities. However, the executive preferred to postpone the difficult decisions suggested by the council. Emmanuel Macron did not expect such a resurgence of the epidemic and was aiming for the country to learn how to "live with the virus."[61]

The end of September nevertheless brings good news: researchers at the Institut Pasteur in Lille have found out a molecule that considerably slows down the growth of SARS-CoV-2 and therefore makes it possible to treat Covid-19. A discovery made possible thanks to the "chemical library" of the Institut Pasteur, a collection of some two thousand drugs marketed around the world. This molecule is therefore already known, as are its side effects, and used as a drug in several countries. The next step of the project is to obtain authorisation for a clinical trial, to recruit patients and to evaluate the effect of the product on patients who have contracted the disease. Researchers hope to start the first human trials in late 2020. The news arrives as the vaccine race rages on in many countries.

Bad news, however, arrives on October 1, 2020. The government decides to place Paris on maximum alert from Monday, October 5, and the indicators are red in several large cities such as Marseille, Lyon, Aix-en-Provence, Grenoble, Saint-Etienne, Lille and Toulouse. That evening, Olivier Véran paints a gloomy picture as Paris exceeds three critical thresholds the day before: the incidence rate of the general population is over 250 cases per 100 000 (exactly 263), the incidence of vulnerable people reached 105 per 100 000 and the share of Covid-19 patients in ICUs rose to 35%. The government concludes that the epidemic is accelerating. It looks like we are back to the March period. Once again, the signals showing that the number of cases is exponentially growing had not been taken seriously. The government is about to shut down bars and restaurants in the capital and other towns, but will that be enough? Olivier Véran and Jérôme Salomon repeat that they are monitoring the situation day to day and are ready to take the necessary decisions, but is it not too late? The virus began to recirculate during the summer, especially among young people, but two months later the impression is that,

[61] Hecketsweiler C, Lemarié A, Faye O, de Royer S (2020, October 17) Covid-19: la semaine où Macron s'est converti au couvre-feu, Le Monde, https://www.lemonde.fr/politique/article/2020/10/17/Covid-19-la-semaine-ou-macron-s-est-converti-au-couvre-feu_6056373_823448.html.

like in March, no one wanted to take responsibility for drastic measures. As a result, without sufficient restriction measures, the virus is spreading to all age groups. Also, on October 1, the scientific council published a new opinion which underlines "the urgency to act" to regain control of the epidemic while some simulations forecast between 4 000 and 12 000 additional deaths on November 1, a month later. The spectrum of a new lockdown looms on the horizon, which would clearly mean that the government's policy had failed. In the discussions, the French often bring up the case of Sweden: how is this country doing so well without having imposed either lockdown or masks?

On October 13, 2020, a day before the President's television speech, the authorities admit that a second wave has arrived. The situation deteriorates across the country and the indicators are red, in particular the test positivity rate, which is close to 12%, and the number of Covid-19 patients treated in ICUs, which exceeds 1 500, the highest figure since May 2020. On October 14, Emmanuel Macron announced a curfew for nine metropoles, including Paris, for at least four weeks which is quite puzzling: while the government claims to rely on science, curfews are usually not among the actions implemented to manage an epidemic. And all this is organised in a hurry, whereas the preceding weeks could have been used to prepare a "plan B."

The situation is becoming critical almost everywhere in France. On October 26, on *RTL radio*, Jean-François Delfraissy estimates that there are nearly 100 000 new daily cases (however SPF announces 33 417 the next day). He admits to be "surprised by this brutality over the last ten days" and adds: "Many people have not yet realised what awaits us," which is somewhat surprising when we see the evolution of the data since the beginning of September. At the end of October, the outbreak is still on the rise (more than 500 daily deaths, 14 000 new hospitalisations over the last seven days, including 2 000 in ICUs, and a test positivity rate at 18.4%). Some hospitals are approaching saturation and the transfer of Covid-19 patients between regions is being organised. To confirm the seriousness of the situation in France and in Europe, *Le Monde* headlines on its front page on October 28: "The outbreak is now out of control." The situation is very tense and the nervousness of the authorities is palpable. According to a poll published on October 22, 64% of the French population do not trust the government to face the epidemic.[62] On October 28, in his fifth television address since the start of the crisis, Emmanuel Macron announces a new national lockdown which will come into force on October 30 for four weeks, but leaving schools, public services, factories, construction sites and farms open. In order to make

---

[62]Chayet D (2020, October 22) Sondage: les Français approuvent le couvre-feu, Le Figaro, https://www.lefigaro.fr/sciences/sondage-les-francais-approuvent-le-couvre-feu-20201022.

progress in a context dominated by growing civil disobedience to sanitary measures, the President calls on the population to "rely on knowledge and science. Never give in to conspiracy, obscurantism, relativism."

This "light" lockdown is seen as a failure of the government's strategy. Yet, as we have seen, warnings and signals had been issued since July. The scientific council had sent clear messages but in a very—perhaps too—cautious way. France has therefore missed its lockdown exit. From June 22 and the deconfinement's phase 3, there were no really binding measures in place. And it is astonishing that a number of experts said they were surprised by the sudden acceleration of the epidemic. Yet it is common knowledge that when we are undergoing an exponential transmission, at some point the curve bends and the process becomes very fast, and out of control—"sudden". This is the very definition of an exponential curve—and an explosion. The Prime Minister was wrong to say on October 29, at a press conference, "that even scientists were surprised by the speed of the virus spread." On September 25, forecasts from the Institut Pasteur already suggested a possible saturation of ICUs from mid-November.[63] In its opinion of October 26,[64] the scientific council does not exclude the possibility of "having several successive waves during the end of winter/spring 2021" and invites the government "to learn lessons from the relative failure of the "testing-tracing-isolating" strategy during the period from May to September 2020." While isolation appears to be the weak link in the anti-Covid-19 chain, the authorities are wondering about the need to make it compulsory for positive and contact cases, with, if necessary, controls and even sanctions as is the case in some countries. Oddly enough, the scientific council says nothing about it. "This is an avenue that deserves a democratic debate," government spokesman Gabriel Attal says on November 20. Almost one year after the emergence of the new virus...

At the same time what we may call the "remdesivir scandal" erupts. On November 20, 2020, WHO releases a conditional recommendation against the use of this antiviral drug in hospitalised patients, regardless of disease severity. These are the conclusions of an expert group, which examined the results of the Solidarity clinical trial, as well as three other randomised controlled trials. "The evidence [suggests] no important effect on mortality, need for mechanical ventilation, time to clinical improvement, and other

---

[63]Barret AL and Focraud A (2020, October 4) Covid-19: les projections qui inquiètent le gouvernement, Journal du Dimanche, https://www.lejdd.fr/Societe/Sante/Covid-19-les-projections-qui-inquietent-le-gouvernement-3996175.

[64]Note of the Covid-19 scientific council (2020, October 26) Une deuxième vague entraînant une situation sanitaire critique, https://solidarites-sante.gouv.fr/IMG/pdf/note_conseil_scientifique_26_octobre_2020.pdf.

patient-important outcomes."[65] However, thousands of doses have already been ordered, at a high price, by dozens of countries. Gilead, the American laboratory which markets the antiviral, fixed the price of one dose at USD 390 in June 2020 while the cost price would be less than one dollar. France, however, did not take part in the order organised by the European Commission. The transparency commission of the Haute Autorité de Santé (HAS, National Authority for Health) has in fact concluded that the actual benefit was too low to justify a financial contribution by social security. There are precedents here: in 2015, the U.S. Senate criticised Gilead's strategy for the development of its hepatitis C drug, aimed at "maximising profit regardless of human consequences."[66] However, amidst the general rush, reason and wisdom prevailed in France's scientific committees...

In early December 2020, the government announces that the vaccination will be free. Two hundred million doses of the Pfizer/BioNTech vaccine have been ordered. Curiously, it is not the Covid-19 Vaccine Committee that will coordinate the operation but a 71-year-old immunologist, Alain Fischer, who will chair the newly created "Vaccine Strategy Steering Board." France is clearly a committee-freak nation: it has created a committee on the vaccine itself, it is now creating one on vaccination... However, Alain Fischer will slow down the vaccination campaign in France, with the support of the President. The (unofficial) reason is that no scientific publication exists yet about the Pfizer/BioNTech vaccine. Actually, they know French people are one of the most reluctant populations in the world to take vaccines. On December 27, 2020, France starts its vaccination campaign at the same time as the other EU countries. But the speed of the roll-out will immediately exasperate the medical staff: as of January 2, 2021, only 332 French people had received the jab while the Covid-19 death toll topped 70,000.

Ten days later, the government is considering imposing a third lockdown. "The quicker you make a decision, the more effective it is," warns Jean-François Delfraissy on January 23, 2021. "A new lockdown will probably be necessary. It's up to politicians to decide. We are in a critical week." But two days later, the President of the scientific council changes his mind: "We are not within a week." Was the information reframed by the Prime Minister? On her side, the President of HAS, Dominique Le Guludec dallies

---

[65]WHO (2020, November 20) WHO recommends against the use of remdesivir in Covid-19 patients, https://www.who.int/news-room/feature-stories/detail/who-recommends-against-the-use-of-remdesivir-in-Covid-19-patients.

[66]U.S. Senate Committee on Finance (2015, December 1) Wyden-Grassley Sovaldi Investigation Finds Revenue-Driven Pricing Strategy Behind $84 000 Hepatitis Drug, https://www.finance.senate.gov/ranking-members-news/wyden-grassley-sovaldi-investigation-finds-revenue-driven-pricing-strategy-behind-84-000-hepatitis-drug.

and waits to know the evolution of the situation as well as the proportion of the so-called British variant amongst the positive tests. "I don't think the government knows today," she says. Again, the lack of decisiveness is striking although the reproduction factor R_t is "significantly" greater than one in several regions. Everyone should know by now that in an exponential phase, when the situation gets worse, it quickly gets out of control…

The situation deteriorates. The number of daily new cases continues to increase steadily, from 10,500 on December 1, 2020, to 25,500 on March 15, 2021. On March 18, the government imposes a light lockdown on Paris and fifteen other departments (so light that the government did not use this word) for a minimum of 4 weeks while keeping schools and essential shops open. At the press conference on that day, the Prime Minister, Jean Castex, does not use science to back up his decisions anymore: "We have taken an approach that is pragmatic, proportional and regionalised." The government most likely regrets not having taken Jean-François Delfraissy's comments of January 23 more seriously and especially not having taken a decision at that time. Because the message of the Chairman of the Scientific Council was then crystal clear and on point: "We realise that if we continue without doing anything more, we will find ourselves in an extremely difficult situation from mid-March." Indeed.

Finally, on 31 March 2021, Emmanuel Macron imposes a third full lockdown for the whole country and closes schools for at least four weeks. It took then two months for the President to follow the advice of his scientific council. Nevertheless, the President claims, for the eleventh time since April 13, 2020, that he now sees the "the light at the end of the tunnel" and the "return of happy days."[67]

London and the "Covexit" Nightmare

The United Kingdom announces its first two Covid-19 cases on January 31, 2020, a week after France. But it is only on March 2 that Boris Johnson chairs the first emergency meeting devoted to the coronavirus, as the outbreak rages on the continent. From the outset, on the response to the coronavirus as well as on many other subjects, Boris Johnson distances himself from the European Union. Is it the Brexit atmosphere that is blowing new ideas

---

[67] France Info (2021, April 2) Covid-19: ces onze fois où l'exécutif a évoqué le « bout du tunnel» et le retour « des jours heureux» depuis un an, https://www.francetvinfo.fr/sante/maladie/coronavirus/Covid-19-ces-onze-fois-ou-l-executif-a-evoque-le-bout-du-tunnel-et-le-retour-des-jours-heureux-depuis-un-an_4357075.html.

around Westminster and giving wings to the government? Or is the emerging epidemic just one more opportunity to assert this very British "difference?" In any case, the Prime Minister claims from the outset that the UK is aiming for coronavirus "herd immunity" and is in favour of the "free circulation" of the virus—quite amusing as the country in the process of leaving the Union. You don't have to be an expert to understand that this strategy may be risky as it involves losing control of the epidemic and assumes that the country's hospital system is robust enough to withstand waves of infections and severe cases.

In fact, across the Channel, the first weeks of the Covid-19 epidemic show that the British government is conspicuous by the absence of a thoughtful and convincing strategy. "The early meetings with the prime minister were dreadful. The general view was it is just hysteria. It was just like a flu, one senior politician says."[68] However, this is not due to a lack of warning. Chris Whitty, Chief Medical Officer for England and the UK government's Chief Medical Adviser, had warned that if the coronavirus got out of China, it would become global, and be on its way to the UK. But the Prime Minister and the government are reluctant to consider anything as draconian as a lock-down. Immediately criticisms were coming in, sometimes virulently from all sides, including from the scientific community, healthcare workers and the media.

On March 2, 2020, the Scientific Pandemic Influenza Group on Modelling (SPI-M) publishes a "consensus statement."[69] The experts, the best known of which are Neil Ferguson of Imperial College London and John Edmunds of the School of Hygiene and Tropical Medicine, are used to advising the government on scientific issues relating to the response to influenza pandemics and other emerging human infectious diseases. They warn authorities that "It is almost certain that there will be sustained transmission in the UK in the coming weeks." However, it transpires from the meeting's minutes that the experts do not have much information at that time: "The critical issue is whether R_0 is greater than one." Government advisers therefore qualify the risk level as "moderate," only briefly considering the possibility of a major outbreak. The next day, March 3, the Prime Minister is his cheery self, shaking hands with everyone and telling the press: "Our country is very well prepared," while Italy has 79 deaths. He adds:

---

[68]Kuenssberg L (2021, March 17) Covid: The inside story of the government's battle against the virus, BBC News, https://www.bbc.com/news/uk-politics-56361599.

[69]SPI-M–O (2020, March 2) Consensus Statement on 2019 Novel Coronavirus (Covid-19), https://assets.publishing.service.gov.uk/government/uploads/system/uploads/attachment_data/file/887541/01-spi-m-o-consensus-statement-on-2019-novel-coronavirus-_Covid-19_.pdf.

"We have a fantastic NHS, the national public health service, fantastic testing systems and fantastic surveillance of the spread of disease".[70]

On March 7, the government advises people showing Covid-19 symptoms to self-isolate, but does not give any sign that the executive is really taking control of the crisis. Five days later, as Emmanuel Macron announces the closure of schools, Boris Johnson continues to advocate for a *laissez-faire* strategy. During a disastrous press conference, the Prime Minister brushes aside criticism and recommends only regular hand washing. However, the day before, Anthony Costello, a leading medical authority in the United Kingdom and a former director at WHO, announces that social distancing could cut the death toll by ten. Patrick Vallance, the Government's Chief Scientific Adviser, seems to keep a close watch and reassures the press: "The UK [is] four weeks behind Italy in terms of the scale of the outbreak."[71] At that time, the experts believe the United Kingdom will handle the epidemic better than the rest of Europe. The next months will prove the contrary.

However, despite government optimism, the crisis looks imminent: during that night, March 7, Chris Whitty announces that the risk for the population of contracting Covid-19 has been re-evaluated from "moderate" to "high." Chris Whitty does not give much detail, but on that day the government receives another epidemiological modelling report compiled by Neil Ferguson's team. This study will not be made public until March 16.[72] In essence, the report says that the *laissez-faire* strategy—the bet on herd immunity hitherto adopted by the Prime Minister—is no longer tenable, given the latest developments in the epidemic and the capacity of the hospital system. At the very least, according to the study, Great Britain could expect in the order of 250 000 deaths. Having updated its model with the latest data, researchers indicate that the NHS would soon be overwhelmed with severe Covid-19 cases and the country might be facing more than 500,000 deaths if the government takes no non-pharmaceutical interventions (NPIs, i.e., public health measures aimed at reducing contact rates in the population and thereby reducing transmission of the virus). The same model suggests that, with an unmitigated epidemic, the United States might face 2.2 million deaths.

---

[70] The British hospital system, which since the end of World War II has provided free healthcare to everyone regardless of their income, still commands enormous respect in the UK.

[71] Cuffe R (2020, March 13) Coronavirus: Three reasons why the UK might not look like Italy, BBC News, https://www.bbc.com/news/uk-51858987.

[72] Ferguson NM et al. (2020, March 16) Impact of non-pharmaceutical interventions (NPIs) to reduce Covid-19 mortality and healthcare demand, Imperial College London, https://doi.org/10.25561/77482.

In British government circles, these computer models are criticised, some emphasising (quite rightly) the decisive influence of the choice of the parameters and their value over the conclusions obtained. The reaction is more surprising when it comes from politicians and economists, whom we know are keen to exploit or at least report on "scientific" forecasts and simulations. However, as of March 12, Boris Johnson and his advisers change their minds. They realise the urgency, and that the official assumptions about the speed of the spread of this new disease have been wrong. But this is not yet made public. We now know that Neil Ferguson's simulations influenced not only Boris Johnson, but also Emmanuel Macron and Donald Trump, and the strategies they are constructing at that time to deal with Covid-19. But it is not in the public domain. Ironically, after having briefed officials in London's Downing Street, Neil Ferguson announces on Twitter on March 18, 2020 that he has a fever and a cough. The mathematical epidemiologist tests positive. He had become a data point in his own study…

Over the following days, a change in attitude is noticeable. The government imposes increasingly restrictive measures, especially for old people, but still no lockdown, despite WHO recommending that everyone of all ages practice social distancing. On March 13, during a press conference at 10 Downing Street, Boris Johnson declares that the government is relying "on science" to respond to the crisis, thus echoing the words of Emmanuel Macron who, exactly the day before, on television, was speaking of his "confidence in science." However, at that time, the scientific information provided to the government has still not been made public.

On March 14, the government's strategy is destabilised by some two hundred scientists and medical professionals who publishes an open letter to the Prime Minister (the letter will receive more than 500 signatures in the following days).[73] Their message is very clear: "Under unconstrained growth, this outbreak will affect millions of people in the next few weeks. This will most probably put the NHS at serious risk of not being able to cope with the flow of patients needing intensive care, as the number of ICU beds in the UK is not larger than that available in other neighbouring countries with a similar population." The scientists are calling for stricter social distancing measures than those currently recommended by the government. On the same day, in Bath, the Half Marathon still attracts more than six thousand runners…

The next day, to everyone's surprise, the government radically changes its strategy. On March 16, Boris Johnson urged British citizens to avoid all unnecessary contacts and travel by promoting teleworking, and staying away

---

[73] http://maths.qmul.ac.uk/~vnicosia/UK_scientists_statement_on_coronavirus_measures.pdf.

from pubs and theatres. The aim is to limit the spread of the virus which has now made more than 50 deaths in one day in the country. Has the experts' warning paid off? The simulations were being shown to be correct despite the fact that Imperial College's work was criticised, with some arguing that their model tends to overestimate the number of hospital admissions, hence the number of deaths. Ferguson himself makes no secret of this fact, but he points out that the main findings of the March 16 report are consistent with what other scientists have found. As with the weather, we know that forecasts are not always accurate. And as with the weather, where meteorology (short-term) and climatology (long-term) are different scientific areas, there are two forms of epidemiological modelling: "nowcasting" (predicting today and the next few days) and "forecasting" (predicting the longer-term future). The second form should be considered with caution because, by itself, mathematics cannot predict the future. On the other hand, they make it possible to simulate and compare different conceivable scenarios.

At the March 19, 2020 press conference, Ian Dunt, editor of politics.co.uk, is met with a tired yet generous Boris Johnson talking about the epidemic. "I don't propose to spend a very long time at this particular one," he tells reporters.[74] "I don't want to weary you with these occasions." The Prime Minister puts forward an arbitrary timetable, announcing the end of the crisis "in twelve weeks." Was Boris Johnson getting bored of the Covid-19 or just tired because he was, at that point possibly infected by coronavirus? Or both? It is plausible, as he will announce a week later to have contracted the disease. In fact, the government had already decided on its action plan a few days earlier but the time has not yet arrived to share it with the public.

However, two days later, on March 22 (which is Mother's Day in the United Kingdom), the prime Minister sent an alarmist message: "We are only two or three weeks behind Italy. The Italians have a superb health care system. And yet their doctors and nurses have been completely overwhelmed by the demand. Unless we act together, unless we make the heroic and collective national effort to slow the spread—then it is all too likely that our own NHS will be similarly overwhelmed."[75] However, the government only announce partial restrictive measures of the population as letters are being sent to the 1.5 million people in England most at risk of coronavirus to tell them to stay at home for at least twelve weeks (the figure announced to the press three

---

[74] I. Dunt (2020, March 23) Johnson already looks bored of the coronavirus, politics.co.uk, https://www.politics.co.uk/blogs/2020/03/23/johnson-already-looks-bored-of-the-coronavirus/

[75] Cullen S (2020, March 22) The UK is just a couple of weeks behind Italy, Boris Johnson warns, CNN, https://edition.cnn.com/world/live-news/coronavirus-outbreak-03-22-20/h_8ef0630d94586355 5317a0480f46a8b7.

days before). He also says that food and medicines will be delivered to their homes. The Prime Minister asks people not to visit their loved ones on that day (Mother's Day), and to follow social distancing guidelines. The number of people who have died in the United Kingdom with coronavirus rises then to 281, as cases reach 5 683.

At this point, the British have no real understanding of what is about to hit them, despite the fact that their European neighbours were already in lockdown. People interviewed on the BBC are "shocked" at the mere idea that they might have to cancel Mother's Day restaurant bookings let alone not be allowed to see their family at all for a number of months. Chris Whitty is talking about "not going too early" as the government assumes that the public would only comply with restrictive measures if they could really feel and see the reasons behind them i.e., hospitals overflowing and loved ones infected etc. The government felt that the public would only tolerate lockdown for a short period of time. In the end, this period has turned out to be much longer than they ever imagined.

The next day, the government changes its position once again. In a short statement pre-recorded at 10 Downing Street, the Prime Minister announces on the evening of March 23 that he has enacted a national lockdown (without using this word) for at least three weeks in order to curb the spread of the pandemic: "From this evening I must give the British people a very simple instruction—you must stay at home." In fact, the Prime Minister already knows at that very moment that the lockdown would extend until mid-June. These are the "twelve weeks" he alluded to a couple of days before. But he is careful not to announce it bluntly at this stage so as not to create panic and general discouragement.

As part of the immediate efforts to contain the virus, the British are only allowed to leave their homes for very limited purposes such as shopping, going to work, receiving medical treatment or exercising once a day, Boris Johnson specifies during the address to the nation. Gatherings of more than two people are prohibited and all shops selling non-essential goods, as well as places of worship, are closed to the public. According to a YouGov poll published the next day, Britons massively support (93%) the Prime Minister's decision to strictly limit people leaving home.[76]

---

[76]The question was the following: "Boris Johnson has announced that, due to the Covid-19 (coronavirus) outbreak, Britons are now only allowed to leave their house in order to: shop for basic necessities; exercise once a day; seek help for a medical need; provide medical care or help a vulnerable person; and travel to work in circumstances where it is absolutely necessary and cannot be done from home. Do you support or oppose these measures?" https://yougov.co.uk/topics/health/survey-results/daily/2020/03/24/43068/1.

Thus, the British government decided to go into total lockdown seven and a half weeks after the first cases appeared on their shores, like France, that is four weeks more than our Chinese benchmark. And it took 21 days for the government to take action after having received the SPI-M report, a delay which, according to the *Times* investigation, resulted in multiplying by a thousand the number of cases and deaths in the UK.[77] The government's delay is all the more incomprehensible since, by mid-March, the epidemic had become pandemic with more than fifty countries in lockdown. Many media outlets and observers point out the indecision and lack of vision of the Prime Minister, but it is more than likely that the economic arguments weighed heavily in the balance and delayed the decision of the executive, which was inevitable. In reality, the delay is even more incomprehensible, for, as we will see, another note was circulating in government circles at the end of February…

Boris Johnson says he took the decision to put the entire country into lockdown on the advice of scientists. He moved from "mitigation" of the epidemic to a strategy of "suppression." This is a significant step-up: while mitigation is an attempt to slow the spread of the outbreak, suppression aims to reverse the epidemic growth altogether by reducing case numbers and keeping them down. The change was announced on March 16. But, in reality, the government had most probably decided to go into lockdown at least a week earlier but the Prime Minister wanted to move there gradually. Indeed, this announcement triggers an impressive mobilisation. On March 25, the government recruited more than half a million volunteers to help the NHS support the most vulnerable people, who must observe strict quarantine, due to their age (70 years and over) or underlying health conditions. The helpers are needed for delivering food and medicines, driving patients to appointments and phoning the isolated.

But on March 26, just three days after the Prime Minister's lockdown announcement, the situation became critical in London hospitals. After having "massively" increased the emergency care capacity in recent weeks, clinics in the capital are facing a "continuous tsunami" and are struggling with the explosion of demand in seriously ill patients. It is therefore more than likely that it was the signals from these hospitals that pushed the Prime Minister to move forward. There is hardly any doubt, as we will see, that he had been fully aware of the peril which his country is going to face for several

---

[77]Calvert J, Arbuthnott G, Leake J and Gadher D (2020, May 23) 22 days of dither and delay on coronavirus that cost thousands of British lives, The Times, https://www.thetimes.co.uk/edition/news/three-weeks-of-dither-and-delay-on-coronavirus-that-cost-thousands-of-british-lives-05sjvwv7g.

weeks. The situation is critical because some hospitals are also facing up to 50% staff sickness rates.

By Skype, Martin Bauer, a social scientist from the prestigious London School of Economics, explained to me on March 26 that his colleagues in the health policy department have been working on this subject for several years: "The liberal system in force in the United Kingdom pushed the NHS to be very competitive and quite cheap compared to its main competitors. In particular, a study carried out by our team has shown that NHS hospitals located in a region with a high hospital supply manage to reduce their operating budget while increasing their efficiency, officially without negative consequences for patients.[78] But it also means that the system has been operating on very narrow margins for years. I am convinced that the capacity of the NHS will be severely tested in the coming days (beds, reanimation devices, tests, protective equipment etc.); it is already a political time bomb."

The sequence of the events in the UK, the conflicting statements from the Prime Minister and the delay in taking action by the government raise questions about the role of experts and scientists in the crisis. In principle, Boris Johnson is well advised and surrounded by top-level scientists. He relies on a group of independent experts supposed to assist the government during emergencies (the so-called Sage committee or Scientific Advisory Group for Emergencies) and has sought their advice on the Covid-19 crisis. However, at that time, Sage looks like a black box: its composition is secret and the minutes and documents of the meetings are not public. It is only on May 29, 2020 that the government decides to publish the minutes of the meetings number 1 to 34 "given the exceptional nature of the Covid-19 pandemic" and to ensure "full transparency on how science advice is being formulated."[79]

The Sage committee reports directly to the Prime Minister and the government's crisis committee, COBRA (from "Cabinet Office Briefing Rooms," which are meeting rooms in the Cabinet Office in London used by the committee co-ordinating the government actions in response to a major crisis). Another key committee is NERVTAG, the New and Emerging Respiratory Virus Threats Advisory Group, of which Neil Ferguson is a member, which advises the Chief Medical Officer and the government on the threat posed by new and emerging respiratory viruses and on options for their management. Boris Johnson can also count on the government's Chief Scientific Adviser, Sir Patrick Vallance, who has a solid curriculum vitae and has

---

[78]Cooper Z et al. (2010, June) Does Hospital Competition Improve Efficiency? An Analysis of the Recent Market-Based Reforms to the English NHS, LSE, http://cep.lse.ac.uk/_new/publications/abs tract.asp?index=3620.

[79]https://www.gov.uk/government/organisations/scientific-advisory-group-for-emergencies.

notably directed the research and development of the global pharmaceutical giant GlaxoSmithKline (GSK). Sir Patrick was of course aware of the work of leading epidemiologists such as Neil Ferguson and John Edmunds (who are also members of the Sage committee). Of course, the central figure here is Boris Johnson and it is well known that the British Prime Minister, often confusing even to his close circle, is reluctant to show any sign of weakness. We have seen this in particular in his very much criticised management of Brexit (Fig. 3.3).

On April 2, I contact Jonathan Leake, Head of Science at the Sunday Times in London, to get his analysis of the current situation. I have known Jonathan for many years because he is a very good connoisseur of the scientific community and British society. He explains: "What I have seen over the past few weeks is that politicians only half-listen to scientists and don't understand the seriousness of the messages. The Prime Minister and his government colleagues were alerted in February to the risk posed by the pandemic and of the restrictive measures that should have been taken as preventive actions. But the government has been slow to take action and the lockdown in place is still too loose, so the epidemic will continue to spread in the UK."

Fig. 3.3 Prime Minister Boris Johnson, Chris Whitty (left), Chief Medical Officer for England and the UK government's Chief Medical Adviser, and Sir Patrick Vallance (right), Chief Scientific Adviser to the Government, at a press conference on the novel coronavirus in London, March 3, 2020 (© Reuters)

I understand that my interlocutor is confused by the inconsistencies of the government. I am also surprised because the United Kingdom has a good reputation and a long tradition in the field of science communication and science-society interactions. Jonathan then mentions a scientific and still confidential note which was sent on February 26, 2020, so more than two weeks before the March 12 study, to the Ministry of Health and Social Assistance as well as to other services which advise the government. This note, which Jonathan kindly sent me and asked me to treat as confidential, is actually a draft scientific publication which presents and discusses a series of simulations of the development of the epidemic according to various restrictive measures, such as the closure of schools, social distancing, protection of elderly people, isolation of severe cases and total lockdown. The conclusion of the authors, almost all from the London School of Hygiene and Tropical Medicine, is very clear: only strict restrictive measures can allow the NHS to withstand the admission of Covid-19 patients and treat the serious cases, and only if these measures are implemented for a "significant duration" within the current year. Quite prophetically, the note predicts a series of waves of infection that would follow one another until Christmas. A minimum lockdown of three months (the famous twelve weeks!) is recommended. The conclusions are not radically different from the Ferguson report. The authors report in particular that an unmitigated Covid-19 epidemic would result in a projected 370 000 deaths directly attributable to Covid-19. The article was finally published in June 2020 in *The Lancet*.[80]

The British government was therefore informed of the risks, the various options and the possible consequences of the Covid-19 outbreak a month before its decision to place the whole country into lockdown. Boris Johnson probably already knew by then what was going to happen but, for reasons one can only guess, was unwilling to jump ahead let alone communicate to the public about it. A delay which, as we have seen, comes at a high price. Boris Johnson must have hoped that the initial note would remain confidential...

At the end of March, the number of contaminations and deaths explodes across the country; the death toll in hospitals is approaching two thousand people and is increasing exponentially. Boris Johnson, who was hospitalised for ten days for symptoms of Covid-19, is criticised more and more in the media, who point out his lack of preparation. Tests, ventilators, protective equipment, basic antibiotics: the essentials are lacking in hospitals and

---

[80] Davies NG et al. (2020, July 1) Effects of non-pharmaceutical interventions on Covid-19 cases, deaths, and demand for hospital services in the UK: a modelling study, The Lancet, vol. 5, issue 7, E375-E385, https://doi.org/10.1016/S2468-2667(20)30133-X.

deaths among healthcare workers are soaring. *The New York Times* head-line on March 26 cruelly sums it up: "Boris Johnson is not cut out for this crisis. Britain needs a leader, not a joker."[81] Controversy is developing in the UK about the infamous eleven-day delay from March 12 to 23, before the government officially abandoned its herd immunity strategy in favour of the lockdown. The government rejects any suggestion that there was a delay, Patrick Vallance explains, adding that the country was almost four weeks behind Italy.[82] On April 12, at the daily press conference in 10 Downing Street, the Secretary of State for Health and Social Care Matt Hancock, who also contracted Covid-19, claims that the NHS has never been overwhelmed. Yvonne Doyle, Director of Health Protection for Public Health England, recommends caution in analysing statistics and comparing them with other countries. However, the NHS continues to lack protective equipment (masks, etc.). A photo of nurses in waste bag gowns is going… viral. The government refuses to admit any mistake in the handling the crisis.

An article published on April 7, 2020 by Reuters, based on interviews with some twenty scientific and medical advisers to the administration and the government, confirms that they had given the alert very early, at the very beginning of March.[83] The article reveals, however, that, in retrospect, these experts acknowledge that they were probably not firm enough, probably so as not to create panic and also because, according to some of them, the population would never have accepted containment measures like those applied in China. I then discover an interview given on March 12 on the *BBC*'s "Newsnight" program by Professor Graham Medley, chairman of the SPI-M modelling committee, member of Sage and one of the government's scientific advisers. In this sequence, which lasts about ten minutes, Graham Medley is particularly confused and hesitant. He says he still hopes for a "good epidemic," which could provide immunity to a large part of the population. The expert has obviously not yet ruled out the possibility that Covid-19 could be treated like flu. Then, he goes on into a rather nebulous discussion about the link between infections and deaths which leads him to say that the change

[81] Russell J (2020, March 26) Boris Johnson Is Not Cut Out for This Crisis. Britain needs a leader, not a joker, The New York Times, https://www.nytimes.com/2020/03/26/opinion/coronavirus-uk-boris-johnson.html.

[82] Cécile Ducourtieux, journalist at *Le Monde* based in London, explains to me by email on April 16 that the government compares the dates on which the United Kingdom and Italy have exceeded fifty Covid-19 deaths.

[83] Grey S and MacAskill A (2020, April 7) Special Report: Johnson listened to his scientists about coronavirus - but they were slow to sound the alarm, Reuters, https://www.reuters.com/article/us-health-coronavirus-britain-path-speci/special-report-johnson-listened-to-his-scientists-about-coronavirus-but-they-were-slow-to-sound-the-alarm-idUSKBN21P1VF.

in behaviour of the population might arrive too late…[84] When Chris Whitty, himself an epidemiologist, presents at the press conference on March 16, 2020, alongside Boris Johnson, the study completed by Neil Ferguson's team, he is keen to temper the findings, saying that the number of infections will likely be "much lower" and that any estimate of the number of deaths is, at this stage, "hazardous." Like in France, advisors and experts, who are among the most competent and the best informed, tend to minimise the gravity of the crisis. Afraid of making a mistake? Afraid to tell the truth?So, what went wrong in the management of the outbreak in the United Kingdom? The country has some of the most talented researchers studying SARS-CoV-2 and tackling Covid-19. But there was a collective failure that prevented the scientific and medical community from taking into account the signals coming from China and Italy. The decision-makers, whoever they are, have obviously not learned the lessons from what is happening in other countries. Weeks have been wasted, resulting in thousands more deaths. Superficially at least, the proximity of scientists and politicians is evident: during press conferences, ministers appear most of the time flanked by their scientific and medical advisers. Has cohesion given way to collusion? Is this proximity only a facade and is it only intended to save face so that the government, and therefore the political mandates that go with it, survive? Does a tacit "give and take" agreement bind all those concerned? Is it an "I give you my credibility" in exchange for an implicit "You ensure my visibility"?

On April 9, 2020, Richard Horton, editor-in-chief of *The Lancet*, claims in The Guardian that the global response to SARS-CoV-2 is the biggest science policy failure in a generation.[85] He does not hesitate to blame researchers and experts. While China, traumatised by the SARS experience, correctly estimated the dangerousness of the new coronavirus and quickly quarantined entire cities, British scientists were still thinking that Covid-19 was a kind of flu. "Our scientists, Horton writes, suffered from a 'cognitive bias' towards the milder threat of influenza." Experts and members of the government failed to complete a proper risk assessment together, which left the NHS unprepared to ride the wave. Scientists and doctors have been unable to learn from each other and to implement the lessons learned from past experiences. "The assessment, in many sectors of government, was that the resulting medicine was so strong that it would be spat out," writes Ian Boyd, a former Chief Scientific Adviser (between 2012 and 2019). "We were poorly prepared, he

---

[84] https://www.youtube.com/watch?v=blkDulsgh3Q.

[85] Horton R (2020, April 9) Coronavirus is the greatest global science policy failure in a generation, The Guardian, https://www.theguardian.com/commentisfree/2020/apr/09/deadly-virus-britain-failed-prepare-mers-sars-ebola-coronavirus.

concludes."[86] However, on April 17, 2020, Patrick Vallance, who, as you will remember, was the head of research and development of GSK, created the "UK Vaccine Taskforce" in order to develop vaccines as quickly as possible for the British population. This turns out to be an excellent decision, to the credit of the British scientific community. The new organism will play a decisive role in the successful vaccine developed by AstraZeneca. By taking risks, the task force also enabled the early conclusion of pre-purchase orders of vaccine doses for the benefit of the United Kingdom.

Another controversy is developing on the counting of deaths. Why, ask the British media, does the United Kingdom not take into account those deaths which have occurred at home or in care homes across the country? Is this a political tactic to downplay the gravity of the situation? Or a logistical impossibility—hardly understandable in fact? It is true that a comprehensive counting would reveal a toll as calamitous as in France or even the highest casualty rate in Europe. But the government refuses to admit it. The situation in care homes is indeed worrying. On April 15, 2020, Matt Hancock maintains his accounting system while promising mass tests for staff and residents in nursing homes.

On April 20, 2020, the first meetings take place, in Downing Street and Westminster, on plans to ease lockdown measures. Several ministers are calling for economic recovery and the opening of schools from May 11, as in France. Their main argument is that the majority of businesses do not have enough cash flow to last more than three months; the economic outlook is dire for the whole country. However, who is ready to take responsibility for several thousand more deaths—and at what cost[87]? In hospitals, the situation is still critical: the NHS has suspended routine appointments and treatments since the end of March in order to admit priority Covid-19 patients, leaving many people on waiting lists, in particular those cancer patients who have been hoping for treatment for several weeks. The government fails to show strong leadership—to say the least—and the Secretary of State for Health and Social Care makes a fool of himself by proudly promoting a pin with the "CARE" logo aimed at medical professionals. Some officials and managers are quick to grasp the opportunity, pointing out that healthcare workers would probably prefer a good salary review and decent working conditions.

---

[86]Boyd I (2020, March 30) We practised for a pandemic, but didn't brace, Nature, https://www.nature.com/articles/d41586-020-00919-3.

[87]Shipman T and Wheeler C (2020, April 19) PM's lockdown dilemma—risk killing the economy or thousands of people?, The Times, https://www.thetimes.co.uk/article/pms-lockdown-dilemma-risk-killing-the-economy-or-thousands-of-people-zk6xrlfhr?utm_source=Silverpop.

On April 26, 2020, Boris Johnson comes out of convalescence and returns to Downing Street amidst criticism from all sides over the NHS' lack of resources, as a result of the austerity policies and competition doctrine carried out by successive conservative governments. While public opinion remains in favour of lockdown, the government is increasingly criticised for its handling of the crisis. Most media outlets offer much of the same analysis: the government was ill-prepared, developed a herd immunity strategy based on an influenza epidemic, and did not at all adapt its plans to the new virus. Another point continues to crystallise criticism: the insufficient stocks of protective equipment and the lack of government initiatives to procure additional resources. However, Boris Johnson is the only British politician who maintains public confidence since the beginning of the coronavirus crisis with 51% favourable opinions, a high score that some commentators link directly to his illness and his two-week absence.

In this context, leaks reveal that the United Kingdom missed four European tenders organised for the massive purchase, at a competitive price, of protective equipment, reanimation and laboratory equipment, despite invitations from the European Commission to participate. The news has a very negative effect as all hospitals are facing the shortage. It looks like Brexit has its first victims. Quite surprisingly, Matt Hancock first confirms that the country was involved in this joint call from the European Union. But, given the denials of the Commission's spokesman in Brussels, Stefan De Keersmaecker, he then explains that his services missed the deadline. A Commission spokesperson confirmed to me that the British services had had plenty of time to respond. So, it might have been a political decision. At the same time, the country was expecting a large shipment of masks from Turkey, but it arrived a week late and incomplete. The administrative services apparently forgot to check whether this shipment of goods could be authorised by the Turkish government. An incredible "fiasco," according to the conservative *The Daily Telegraph*.

And, still on April 26, another hiccough puts Boris Johnson in an uncomfortable situation: a scoop from *The Guardian* reveals the presence of the Prime Minister's adviser, Dominic Cummings, at several meetings of the Sage scientific committee, including that of March 23, the date on which the lockdown was announced. Of course, Cummings may have participated as an observer but experts says that he regularly intervened in the discussions—information that immediately cast doubt on the real independence of the committee.

Also, and above all, the outbreak's toll is very heavy: the number of Covid-19 deaths exceeds 20 000 on April 26 for the whole country. And the situation worsens on April 28 when the health authorities, under pressure from the media, consolidate the number of deaths in hospitals and care homes for the first time. This makes the country hit a record fortnight of more than 4 400 deaths (between April 10 and 24). From that date onwards, care home deaths are tracked in the same way as hospital fatalities. With 26 097 deaths in total as of April 29, the United Kingdom becomes the fifth country in the world in terms of Covid-19 mortality (after Belgium, the Netherlands, Spain and Italy). We are far from what Sir Patrick Vallance said one month earlier: "If we can get [the number of deaths] down to 20 000 and below that's a good outcome in terms of where we would hope to get to with this outbreak."[88] Great Britain, an island that has barely one percent of the world's population, concentrates twelve percent of total Covid-19 cases. While still defending his counting method two week earlier, Matt Hancock had to change his mind. People realise across the country that the early analysis of the situation had been woefully underestimated and that this was going to be significantly worse than everyone thought. Critics and calls for the resignation of the government immediately burst in the media and on the social networks.

From May 5, 2020, the government is putting all its efforts into promoting the "NHS Covid-19" coronavirus tracking application that Britons will be able to download on their smartphones. Like StopCOVID in France, it uses Bluetooth technology to inform its users of the possible proximity of infected people and, conversely, to help them alert their close contacts if they test positive for coronavirus. Speaking to the public, Matt Hancock says it is everyone's duty to use this app which will "save lives," also reassuring them about the privacy of personal information. Its success will depend on the rate of use of the application, which the government expects to be downloaded by 60% of the population. At the same time, the media announces the resignation of star epidemiologist Neil Ferguson from the Sage committee for breaking lockdown by receiving his lover at his home on multiple occasions. He follows the Scottish government's medical adviser, Dr Catherine Calderwood, who resigned on April 5 for travelling to her second residence, thus breaking the confinement rules she had helped define. It seems that scientists, too, can make their own rules... Ferguson will be continuing to advise the government as an official member of NERVTAG.

On May 10, Boris Johnson announces that lockdown stays in place at least until June 1, 2020, given that the conditions have not yet been met to return to "normal" life. Above all, numbers are still very bad in the UK.

---

[88] Haynes L (2020, March 17) Social distancing could keep UK Covid-19 deaths below 20,000, GP, https://www.gponline.com/social-distancing-keep-uk-Covid-19-deaths-below-20000/article/1677392.

However, in reality, the government surreptitiously starts easing lockdown without being explicit about that. The Prime Minister wants to "beat the virus" and at the same time "reopen society." He encourages Britons who cannot telework to start returning to the workplace. He also agrees to lift some of the strict lockdown rules, for example by allowing from May 13 (and not May 11 as originally announced) meetings outside limited to two people. However, the government's decision to change the slogan from "Stay at Home" to "Stay Alert, Control the Virus," further adds to the confusion. As muddled as the Prime Minister's strategy for tackling the epidemic was, his plan for lockdown exit appears to lack preparation and consistency. During a meeting on May 12 to discuss the plans to ease restrictions, Boris Johnson asks Sir Mark Sedwill, the then acting Cabinet Secretary, who got Covid-19 at the same time as the Prime Minister but without publicly admitting it: "Who is in charge of implementing this delivery plan?" After a silence, the Prime Minister looks at Sir Mark and says: "Is it you?" Sir Mark then reportedly replies: "No, I think it's you, Prime Minister." According to several sources close to the government, no one in the government has a clear idea of what to do and all are avoiding a misstep, ahead of an inevitable public inquiry into the government's handling of the crisis that might be launched at the end of the epidemic. As a result, the British admit to feeling lost by the lack of clarity of the government's communication and the feeling that it is pushing its responsibilities off onto the population.

The government is clearly on the defensive. Some of its members do not hesitate to question the role of the experts who advise them. Sir Adrian Smith, a statistician (who became President of the Royal Society, Britain's most distinguished scientific society, in November 2020), needs no encouragement to put aside his scientific reserve.[89] The scientist blames Boris Johnson's ministers for hiding behind the experts ("We are simply doing what scientists tell us") and asks them to be more open about the advice they receive and to acknowledge the "extraordinary amounts of uncertainty" that remain. He fears that a public in-depth review of Britain's handling of the coronavirus outbreak may draw scientists into a blame game. "It is the politicians who ultimately make the decisions," Sir Adrian recalls. The scientist also points out the lack of openness and transparency on the government's side. True, the list of members of the Sage committee was not published until the beginning of May. True, the minutes of its meetings will be published only on May 29 although at that time only twenty-eight reports out of the 120 produced

---

[89]Whipple T (2020, May 19) Coronavirus: Stop passing the buck, top scientist tells politicians, The Times, https://www.thetimes.co.uk/article/coronavirus-stop-passing-the-buck-top-scientist-tells-politicians-d6292nbrg.

by the Committee will be available on the website. "Even if nothing terribly secretive and terrible is going on, you feed suspicion if you're not transparent," Sir Adrian concludes. His intervention comes at a time when many Members of Parliament are questioning the government about the lack of transparency that characterises the work of its scientific committees.

This whole issue about navigating between what we know and what we don't know is far from specific to the United Kingdom: everywhere, the political world prefers facts, which reinforce decisions, while the current situation is full of unknowns. Unfortunately, the scientific world is sorely lacking in experimental data, not only on the virus itself but also on its effects on humans, on the impact of lockdown, etc. so that we are all part of an extraordinary "global experience" at that time. On May 19, Secretary of State for Work and Pensions Therese Coffey said the government had received "bad" scientific advice, which she said was responsible for the high mortality in care homes. Despite these criticisms, Sir Adrian told me by email on May 20 that he does not believe that a "scientific witch hunt" could take place in his country, which has a long tradition of scientific culture.

However, it is a non-scientific affair that is going to embarrass the Prime Minister over the next days: on May 22, *The Daily Mirror* and *The Guardian* reveal that the government's adviser Dominic Cummings left his London home at the end of March, in full lockdown, to visit his parents in Durham, in the north-east of England, despite himself having had symptoms of Covid-19. Immediately the Labour Party, sitting in opposition to the government, condemns a two-speed regime ("There cannot be one rule for Boris Johnson's most senior adviser and another for the rest of us") and calls for an urgent inquiry into allegations that Boris Johnson's key adviser broke coronavirus lockdown measures… The controversy, which has become political, will capture the attention of the media and incite the anger of its readership for several weeks. The public indignation for this hypocrisy was indeed massive. The British are obedient in following the rules but completely lacking in grace for anyone who slips up. As a result, Cummings (as well as Ferguson) were thoroughly destroyed in the media.

On June 3, the United Kingdom hits at record-high of 40,000 deaths as the virus transmission rate, the famous R_t, remains close to 1. However, in the House of Commons, Boris Johnson cries victory, responding to the opposition: "If you look at what we have achieved so far, it is very considerable. We have protected the NHS. We have driven down the death rate. We are now seeing far fewer hospital admissions. I believe that the public understand that, with good British common sense, we will continue to defeat this virus and take this country forward, and what I think the country would like to hear from him is more signs of co-operation in that endeavour." Criticised for going into lockdown too late, Boris Johnson is now facing criticism, including from within his own party, for lifting the lockdown in an inconsistent and rushed manner. For example, quarantining travellers arriving from abroad is only implemented from June 8. Primary schools, which reopened on June 1, have an attendance rate of around 50%. And the NHS Covid-19 contact tracing app is not yet fully operational.

The next week is difficult for the Prime Minister. On June 9, a plan to get primary schools fully open again before the summer is delayed till September. Then, the British Parliament starts evaluating the government's handling of the epidemic. Speaking at the House of Commons' Science and Technology Committee on June 10, Neil Ferguson, whose opinion was instrumental in the lockdown strategy, explains to MPs the essence of his team's work and conclusions. According to him and his fellow modellers, imposing a lockdown a week earlier would have cut the death toll by half: instead of counting at that time more than 40 000 victims, the number would have been around 20 000. Ferguson also criticises the government for failing to protect the elderly living in care homes. Boris Johnson quickly replies that it is too early to make such a statement[90]: "We will have to look back on all of it and learn the lessons that we can. A lot of these things are still premature. This epidemic has a long way to go." Chris Whitty, however, is more open: "There's a long list of things we need to look at very seriously, in particular the UK's slowness to increase its testing capacity at the start of its outbreak. Part of the problem at that stage is that we had very limited information about this virus."[91]

On June 18, the government acknowledges the failure of the NHS Covid-19 application, which, after three months of development, still only works on the Isle of Wight. On June 18, Matt Hancock announces that the application will be replaced by a tool using a technology developed by Apple and Google. The main problem appears to be the fact that NHS Covid-19 only detects 4% of Apple-branded phones in the vicinity of the person using it.

On June 23, Boris Johnson announces the easing of the lockdown in England from July 3. "Our long national hibernation is beginning to come to an end, […] a new, but cautious, optimism is palpable," declares a triumphant Prime Minister in the House of Commons. However, many

---

[90] BBC News (2020, June 10) Coronavirus: 'Earlier lockdown would have halved death toll', https://www.bbc.com/news/health-52995064.

[91] Cooper C (2020, June 10) Scientists turn on Boris Johnson over UK's coronavirus response, Politico, https://www.politico.eu/article/scientists-turn-on-boris-johnson-over-uks-coronavirus-response-Covid-19/.

observers point out that this decision comes too soon. In terms of Covid-19 mortality, the UK is still currently in the top ten: see Fig. 7. With nearly two hundred daily new deaths, the epidemic is still far from under control. And everything seems to concur that this sad record will be soon exceeded. On July 20, simulations show that delays in diagnosing and treating cancer due to the coronavirus could lead in the worst case to up to 35 000 more deaths in a year. It is estimated that as many as two million routine breast, bowel and cervical cancer screenings have been cancelled or postponed due in part to the disruption of health services and the NHS in particular.[92] In some hospitals, radiation therapy machines are in a sleeping mode although they could have been used to save lives.

On July 24, the government imposes the wearing of face masks and coverings in enclosed places, to the great satisfaction of the British Medical Association, which represents the medical professionals. Until then, wearing a mask has only been mandatory on public transport and at NHS facilities, and since June 15 has only been recommended in enclosed public spaces. However, there are still some inconsistencies that are incomprehensible to the public: the mask is mandatory in shops, but not for staff, and not in pubs and restaurants.

On August 13, the government announces a surprising reduction in the number of Covid-19 deaths, which drops from 46,706 to 41,329. The health department has indeed changed its method of counting the deaths attributed to the disease. Until that date, anyone who tests positive for SARS-CoV-2 and who dies is registered as a victim of Covid-19. However, it has been recognised that some of these people would have died from other causes. The deaths are now attributed to Covid-19 if they occur within 28 days of the test, thus lowering the official death toll by nearly 5 000 units.

At the beginning of September, around the same time as in France, a second wave of Covid-19 arrives in the United Kingdom and hits care homes in particular. With nearly 200 000 daily tests performed nationwide, laboratories are overwhelmed, queues are growing and chemical reagents are running low in many regions. Following a doubling of the number of infections each week, the government takes a regional approach and the whole of north-east England is once again under restrictions from September 14. The number of participants in meetings is limited to six people, whether they are held indoors or outdoors; pubs and restaurants must adhere to a strict curfew

---

[92] Sud A et al. (2020, July 20) Effect of delays in the 2-week-wait cancer referral pathway during the Covid-19 pandemic on cancer survival in the UK: a modelling study, The Lancet Oncology, https://doi.org/10.1016/S1470-2045(20)30392-2.

and close at 10 pm, and meetings between different households are prohibited. Barely a month after inviting the British to return to the office, Boris Johnson makes a 180° turn in the House of Commons on September 22: "First, we are once again asking office workers who can work from home to do so." Nearly 9 million Britons are affected by this return to partial lockdown. However, contrary to the first wave, the situation the country is facing was not inevitable. Scientists warned that, without taking urgent action over the summer to mitigate the risks, this is what would happen. But the government did not heed their warnings. The government is seeking at all costs to avoid a second lockdown and the closure of schools. The NHS Covid-19 app goes live nationwide on September 24, and is downloaded twelve million times in just four days.

On September 23, Clive Cookson, head of the Financial Times' science pages, announces that the British authorities are considering inoculating SARS-CoV-2 in tens of thousands of young healthy volunteers who had previously been administered an experimental vaccine in order to observe the effectiveness of the protection.[93] The procedure, which raises serious ethical questions, offers the advantage of shortening and simplifying the lengthy process of testing vaccine candidates. After a few weeks or months, it will suffice to compare the state of health of the volunteers with that of a control group having received a placebo. But the prospect of deliberately infecting people—even those at low risk of severe disease—raise several issues as in particular people may participate for the money without appreciating the risks. The origin of the vaccine candidate has not been disclosed except that it does not belong to AstraZeneca.

Faced with this second wave, Boris Johnson, who was slow to act in March, wants to minimise criticism. He considers his options as the situation deteriorates across the country. On October 12, the government announces a three-tier lockdown system through a regional approach, imposing restrictions locally depending on the severity of the epidemic. Socialising in particular is being curbed further in areas that are being placed in the 'high alert' and 'very high alert' categories. For example, with the new system, meeting friends and family indoors is now banned in London and other parts of England. According to Mayor Sadiq Khan, the infection rate is on a steep upward path in the British capital, with the number of cases detected through NHS Test and Trace (an NHS service that traces the spread of the virus in order to isolate new infections) doubling over the last ten days. The 7-day

---

[93]Cookson C (2020, September 23) UK to test vaccines on volunteers deliberately infected with Covid-19, Financial Times, https://www.ft.com/content/b782f666-6847-4487-986c-56d3f5e46c0b.

average case rate on that day stands at 100 per 100 000 people, and is still rising sharply.

However, as in March, the government's response is not convincing and does not appear to be up to the task. The Prime Minister is criticised on October 13 for having ignored the recommendations of the Sage scientists. According to the report of a meeting which took place on September 21 (but published on October 12), the experts had recommended the immediate establishment of a "short lockdown" to act as a "circuit breaker" and reverse the exponential increase in cases.[94] They also recommended other measures, such as closing all bars, restaurants, indoor gyms and personal services, working at home for all those who can, online-only classes at universities and colleges, and banning in-home contacts with members of other households. For the most part, Boris Johnson simply went with the latter restriction, an approach which Chris Whitty himself considered insufficient. The United Kingdom therefore practically follows France, which is facing a second wave at the same time with regional restrictions and the limitation of private meetings for the former, curfews and the closing of bars for the latter. Despite the "exponential progression" warnings that had been abounding since early September, these two countries once again missed the boat with a tardy and insufficiently broad response. Admittedly, the country is better prepared than in March to face this new wave, especially at the hospital level where there are more reanimation beds and enough protective equipment for the staff, but the image of Boris Johnson is clearly degraded in the political spheres and in public opinion. The Prime Minister has yet again failed to come up with a meaningful anti-coronavirus strategy and Britons believe the government is too lax.

According to a poll carried out on October 15 and 16 for *Sky News*, the television channel, two thirds of the people interviewed say they are ready to accept a brief lockdown in England and 61% no longer trust the Prime Minister to make the right decisions about Covid-19. Sage's member Jeremy Farrar comments: "It's never too late to impose temporary national lockdown but the best time was last month." More worrying politically speaking, Boris Johnson is increasingly criticised in his own camp. Like Emmanuel Macron at the same time, the Prime Minister is being pressed by his own party to put forward a strategy for "living with the virus." Added to this is a growing distrust of mayors towards measures decided by officials in the capital whom they see as being too far from the real world. The soar in popularity Boris Johnson experienced after the first lockdown is a long time ago:

---

[94] https://www.gov.uk/government/publications/fifty-eighth-sage-meeting-on-Covid-19-21-september-2020.

the entire government is now accused of incompetence. Six months later, a senior minister said: "We should have locked down more severely, earlier in the autumn—the whole point was, the earlier you act the more you buy yourself time for a strategy that can get out."[95]

Has Boris Johnson turned his back on science? Although he claimed, like Emmanuel Macron in France, to "rely on science," is he now distancing himself from his scientific advisers? The reality is undoubtedly more prosaic and we will come back to this issue in the next chapter.

On October 31, while the country has more than a million infections, the Prime Minister announces, alongside his scientific and medical advisers, a *re-lockdown* of England from November 5 until December 2, after having previously discussed the situation by videoconference with his ministers. Pressing Boris Johnson to move into lockdown for several weeks, his advisers have this time won the battle—the reasons are mainly the severity of the second wave and the prospect of allowing families to reunite at Christmas. Yet more backpedalling which spectacularly demonstrates the failure of his anti-coronavirus strategy and which further erodes the government's political credibility. The next day, November 6, the city of Liverpool offers its entire population—almost half a million inhabitants—the option to get tested every two weeks. Some 2 000 troops are mobilised to set up mobile test centres throughout the city. However, the key question remains unanswered: what about monitoring people who are supposed to self-isolate, namely those who have tested positive and their close contacts?

On November 10, Matt Hancock announces an ambitious vaccination plan that could be implemented as early as December 1 if the vaccine produced by Pfizer and BioNTech is ready and receives the green light from the national regulatory agency. Operations will be coordinated by the NHS and the military will be called in to deliver the doses—a complicated logistics because the vaccine must be stored at -70 °C. The government is counting on the positive image of the NHS in the public's psyche to limit the influence of anti-vaccine movements. Boris Johnson delayed his first lockdown but he does not want to delay his first vaccination campaign… On November 23, Boris Johnson announces to the House of Commons the end of national lockdown on December 2 and the implementation of a "Winter Plan" proposing three-tiered restriction regional measures "in some ways tougher than the pre-lockdown measures," he adds. The decision comes as the number of deaths is still on the rise and the country is top eighth in the world ranking list of Covid-19 mortality.

---

[95]Kuenssberg L (2021, March 17) Covid: The inside story of the government's battle against the virus, BBC News, https://www.bbc.com/news/uk-politics-56361599.

These few pages may give the impression that Boris Johnson bears a heavy responsibility for the failed handling of the epidemic. He certainly made mistakes, in particular by delaying the first lockdown and by not making mask-wearing in public compulsory before the autumn of 2020, but he also implemented a strategy designed and validated by government experts, all of them internationally renowned scientists. Mismanagement is the sinister responsibility of the entire government, which made bad decisions, based in part on inappropriate advice, itself based on poor quality data. It is therefore the case that the government has collectively failed and must accept responsibility for presiding over one of the world's most serious health crises. Boris Johnson's ministers failed to anticipate, diagnose and manage the crisis. One explanation is that the crisis has sharply exposed the systemic weaknesses of the United Kingdom, which is now paying a heavy price for mistakes made several years ago. The civil·service and many parts of the administration were simply overwhelmed for lack of preparation. The Department of Health could not provide enough tests; the Foreign Office was unable to quickly recall Britons abroad; the NHS was unable to manage and share data on registered patients. The analysis of Sir Mark Walport, one of the members of the Sage committee and a former Managing Director for UK Research and Innovation (UKRI) from 2017 to 2020, is that "many of the challenges that we've had are not, as it were, about policy advice or the science advice; they are questions about resilience."[96] Then the government seemed to be more concerned about the economic recovery than the health crisis.

On December 2, the authorities gave the green light for the BioN-Tech/Pfizer BNT162b2 vaccine of which 40 million doses had already been ordered. Therefore, the United Kingdom becomes the first country to licence a fully tested Covid-19 vaccine, as MHRA, the regulatory agency, gave *temporary* authorisation in one day. A record speed which has been qualified as a "Brexit success." Indeed, in the EU, the approval of the new vaccines has to go through the European Medicines Agency (EMA) and the European Commission.[97] The government intends to launch the vaccination campaign the week after. Indeed, on December 7, 2020, Margaret Keenan, 90, a former jewellery shop assistant, is the first Briton to be vaccinated as part of the country's mass inoculations against Covid-19, and the first person anywhere in the world to

---

[96]McTague T (2020, August 12) How the Pandemic Revealed Britain's National Illness, The Atlantic, https://www.theatlantic.com/international/archive/2020/08/why-britain-failed-coronavirus-pandemic/615166/.

[97]Indeed, the United Kingdom would usually wait for the European Medicines Agency to approve a vaccine before looking to distribute it, but in an emergency, EU countries are allowed to use their own regulator to issue temporary authorisation.

receive a clinically authorised, fully tested coronavirus vaccine. She was given the shot at a hospital in Coventry, in central England.

In the meantime, the sky is darkening again. On December 19, 2020, Boris Johnson announces a relockdown of London and south-east England the following day, in response to a worrying increase in the number of infections. The epidemic does not show any signs of slowing up: on January 9, 2021, Britain reports the highest official toll from Covid-19 in Europe as deaths pass 80 000 and confirmed cases top 3 million. This new increase seems to be linked, at least in part, to the appearance of a new strain of SARS-CoV-2, called VUI-202012/01, which appears to be progressing very rapidly.[98] Based on initial analysis, NERVTAG, the New and Emerging Respiratory Virus Threats Advisory Group which advises the Chief Medical Officer (and of which Neil Ferguson is a member), concludes at its meeting on December 18, 2020 to have "moderate confidence that VUI-202012/01 demonstrates a substantial increase in transmissibility compared to other variants." However, it is not yet sure, at that time, if this variant is associated with any changes in the severity of symptoms, antibody response or vaccine efficacy. This news played a key role in changing the Prime Minister's mind – and led to the cancellation of Christmas plans for millions of Britons. The fact that Boris Johnson had to row back on Christmas plans at a very late stage was devastating for the British, for whom Christmas lunch on December 25, where millions of families across the nation traditionally sit down together to eat a turkey roast, is burned into their national identity! One month later, the variant is found in some thirty countries. And a study published on March 10, 2021 demonstrates that the 22 mutations of this variant increase the risk of mortality by 64% (of 55 000 people infected, the previously circulating variants lead on average to 141 deaths, against 227 for VOC 202,012/01).[99]

And yet more bad news: Pfizer laboratories have decided to move their vaccine packaging and distribution plant in Havant to Puurs in Belgium, because of Brexit. Which thus brings British people back to a harsh reality: they are exposed to the consequences of the "Covexit" as it turns out that the epidemic and the UK's exit from the EU are interfering with each other. Despite a relatively positive start to 2021 (the launch of the vaccination campaign against Covid-19 and the signature of an agreement on Brexit on December 24, 2020), they may have to pay the price of the calamitous

---

[98]Rambaut A et al. (2020, December 19) Preliminary genomic characterisation of an emergent SARS-CoV-2 lineage in the UK defined by a novel set of spike mutations, Virological.org, https://virological.org/t/preliminary-genomic-characterisation-of-an-emergent-sars-cov-2-lineage-in-the-uk-defined-by-a-novel-set-of-spike-mutations/563.

[99]Challen R et al. (2021, March 10) Risk of mortality in patients infected with SARS-CoV-2 variant of concern 202,012/1: matched cohort study, BMJ 2021: 372, https://doi.org/10.1136/bmj.n579.

government handling of both the EU exit and the coronavirus crisis, which will have far-reaching consequences for the United Kingdom. However, early 2021, as vaccination is entering a critical phase, the government presents the success of its vaccination campaign as a happy side effect of Brexit. AstraZeneca is meeting its doses contracted for the United Kingdom but has missed its target for the EU, sparking tensions over the channel.

After January 20, 2021, the number of infections is significantly decreasing, leading Boris Johnson to announce one month later, on February 22, a four-step roadmap out of lockdown which may end on June 21, 2021 at the earliest with the removal of all limits on social contacts and a complete reopening of the economy. On 2021, March 28, Public Health England figures show no deaths had been registered in London of patients within 28 days of a positive coronavirus test.

In Washington, The "Mask Politics"

On the other side of the Atlantic, the story of how Covid-19 traced its way through the United States will remain one of the great paradoxes of the pandemic. No other country was perhaps better equipped to tackle this crisis from a scientific, technical and industrial point of view. And yet, in about one year, more Americans have died from Covid-19 than from World War II, the Vietnam War and 9/11 combined.

Of course, the pandemic developed in a particular public health and political context, but this does not explain everything. In terms of health care, Uncle Sam's country is characterised by a relatively complex and inefficient system, and aging infrastructures. The United States remains the only industrialised nation that does not have a universal health care system. Public insurance only covers certain categories of people (elderly and/or disadvantaged), which explains why the method of reimbursing medical expenses is mainly based on private insurance and that around 15% of Americans do not have any health insurance. Unsurprisingly, the crisis has exposed the dysfunctionality of a health system that leaves nearly 30 million people behind. On the political front, in early February 2020, the President emerges triumphant from an impeachment trial that saw Democrats and Republicans clash for several months. Reinforced by this relative political success, Donald Trump is in a strong position, crowned by the U.S. economic performance. With an annual growth of more than 3% and an unemployment rate at the lowest level since the end of the sixties, the President is keen to claim in his third State of the Union address on 5 February 2020 that the economy is "the best it has ever been." He goes as far as to say the nation is "moving forward" at an "unimaginable" pace—a claim not backed up by the data—"and we will never back down."

The first official Covid-19 case, a 35-year-old man who stayed in Wuhan, is identified near Seattle on January 21, 2020. That day, Nancy Messonnier, director of the CDC's National Center for Immunization and Respiratory Diseases, confirms the information that had been circulating since January 17 and considers the news "concerning." The next day, January 22, Donald Trump is setting the pace: "It's one person coming in from China. We have it under control. It's going to be just fine," the President tells *CNBC* from the World Economic Forum in Davos.

Unfortunately, the course of the epidemic will also become exponential here. However, Donald Trump assures us that the disease will "go away in April" (February 14), "one day it's like a miracle, it will disappear," (February 27) and that "it's a very contagious virus [...] but it's something we have tremendous control of" (March 15). He then calls the disease a "new hoax" invented by the Democrats and the media (February 29).

We will see in the next pages that confidential notes and reports from intelligence agencies informed the President of the severity of the epidemic as early as the end of January, therefore since the very first case. While some mayors are starting to ban public gatherings, Donald Trump is hostile to any restriction measures: "It will pass. Each year, between 27,000 and 70,000 Americans die from seasonal flu. Nothing is shut down, life and the economy go on," he tweets on March 9, 2020.

However, with the increasing number of cases, the President can no longer deny the evidence. He declares, on March 13, the coronavirus pandemic a national emergency, thus allowing the federal budget to provide additional resources up to USD 50 billion to states and localities affected by Covid-19. Following up this declaration, Donald Trump rewrites history in his own way: "I have always known that this is a real—this is a pandemic. I felt it was a pandemic long before it was called a pandemic," he declares on March 17, nearly two months after the first case, during the press conference that takes place that day at the White House. The day after, he presents himself as a "wartime President" and asks for special means, calling on "Every generation of Americans to make shared sacrifices for the good of the nation. In World War Two, young people in their teenage years volunteered to fight. [...] And now it's our time."[100] The President also announces the closure

---

[100]Oprysko C and Luthi S (2020, March 18) Trump labels himself 'a wartime president' combating coronavirus, Politico, https://www.politico.com/news/2020/03/18/trump-administration-self-swab-coronavirus-tests-135590.

of the border with Canada, the U.S.'s second economic partner, except for essential travel and the transit of goods. Many media outlets highlight the chaotic and contradictory speeches of the President on the epidemic to date.

One day later, on March 19, the governor of California orders all 40 million inhabitants to stay at home amid the outbreak. He will be followed the next day by the governors of the states of New York, Illinois, New Jersey and Connecticut.

At this point, Donald Trump is still hesitant about the next steps, despite the surge in the number of cases and the collapse of the Dow Jones, one of his favourite indicators. He knows he is up for re-election in November 2020 but still (or hence) refuses to impose a federal lockdown. He assures in a tweet on March 22 that "in two weeks we will have a discussion on the direction we want to take." On that day, *CNN* claims to have counted 33 false claims made by the President about the coronavirus crisis in the first two weeks of March.

On March 23, the *New York Times International* ignites a fire with a front-page article: "The virus can be stopped, but only with harsh steps."."[101] In short: if the United States does not carry out large-scale testing and impose extreme social distancing, the consequences will be devastating for the population. Behind the scenes at the White House, public health experts and the President's economic advisers are at loggerheads. Has the government gone too far in the restrictions? Or not far enough? Should the country take the same route as the United Kingdom which, on that very day, imposed a total lockdown?

At that time, the city of New York alone accounts for 5% of the total number of confirmed cases. But it is always a kind of cacophony that wins the day. The industrial and financial world, which is starting to panic when seeing the Dow Jones losing 37% of its value since February, speaks through the voice of Donald Trump's economic adviser, Larry Kudlow, on Fox News on March 23: "You have to ask yourself whether the shutdown is doing more harm than good. We cannot shutdown the economy. The cost is too heavy for individuals. More testing is essential, and we're loading up with tests now." He concludes that "We're going to have to make some difficult trade-offs."

That same day, March 23, the mayor of New York, Bill de Blasio, calls for a national lockdown or, more softly, a shelter in place order: "The strict measures taken by New York and California (stopping non-essential activities and the obligation to remain home) should be applied everywhere," he says on *CNN*. In his—almost daily—press conference, Donald Trump speaks

[101] McNeil Jr. DJ (2020, March 23) The Virus Can Be Stopped, but Only With Harsh Steps, Experts Say, https://www.nytimes.com/issue/todayspaper/2020/03/23/todays-new-york-times.

at length about the health problems… of the American economy and the increase in the rate of unemployment. Without deviating from his legendary optimism (or recklessness): "America will again, and soon, be open for business—very soon—a lot sooner than three or four months that somebody was suggesting. A lot sooner. We're not going to let the cure be worse than the problem."[102] On the same day, the Senate and the White House reach an agreement to inject two trillion dollars into the American economy, which is already suffering heavily from the coronavirus: more than 3 million Americans filed for unemployment benefits the previous week, breaking the record of the 2008 financial crisis.

It happens that various personalities will take a stand in favour of lockdown (or shutdown), such as Bill Gates, co-founder of Microsoft and Co-Chair of the Bill and Melinda Gates Foundation. Gates had raised as early as 2015 the risk of a global pandemic similar to that of Covid-19. On March 24, during a TED Connects program broadcast online, he lets the voice of reason be heard[103]: "We did not act fast enough to have an ability to avoid the shutdown. Government officials across the country have advised or directed residents in the past days to stay home in a bid to slow the spread of the coronavirus. Many locations, including California, New York City and Washington, D.C., have ordered all nonessential businesses to temporarily close. As a result, unemployment claims are surging and markets are hitting multiyear lows." He suggests an "extreme" shutdown of six to 10 weeks, adding that "everyone should have been on notice in January."

However, that same day, Donald Trump still rejects the idea of a general shutdown for the country, while sixteen states, representing nearly 58 million inhabitants, decide to close all non-essential activities.

On March 26, New York becomes the epicentre of the epidemic, with the number of deaths surpassing 200 per day and tripling from the previous 24-h period. The situation turns into a nightmare, with overcrowded hospitals and a dire lack of protective equipment, mechanical ventilations and ICU beds. Makeshift morgues and field hospitals are being set up in tents in several parts of the city, including Central Park, in anticipation of the exploding number of deaths. In some establishments, the dead bodies are taken away by refrigerated trucks. However, Donald Trump announces that day that he wants to relax the social distancing measures based on a local risk assessment. In a letter

---

[102] https://www.whitehouse.gov/briefings-statements/remarks-President-trump-vice-President-pence-members-coronavirus-task-force-press-briefing-9/.

[103] Bursztynsky J (2020, March 24) Bill Gates says the U.S. missed its chance to avoid coronavirus shutdown and businesses should stay closed, CNBC, https://www.cnbc.com/2020/03/24/bill-gates-us-missed-its-chance-to-avoid-coronavirus-shutdown.html.

addressed to the governors of the country, he underlines that "the battle will still be long" although he said that he was ready to reopen the economy for Easter a few days before. Some people wonder whether the President ever leaves his house and realises what is happening in his own country.

On March 30, Donald Trump U-turns and extends the social distancing guidelines to April 30, as thirty states urge their residents to stay at home. The projections are catastrophic and indicate that the United States could become the most affected country in the world with as many as 200 000 deaths (which, as we know now, is far below the true figures).

On April 3, at the press conference, a journalist asks Donald Trump if he agrees that all states should require their inhabitants to stay at home, as proposed by his coronavirus adviser Anthony Fauci. Fauci, 80, director of the National Institute of Allergy and Infectious Diseases, has worked with five successive Presidents and has grown to become one of the most trusted authorities on the coronavirus. He is today the reference for the pandemic in the United States. But the President kicks the ball into touch: "I leave it up to the governors. I like that from the standpoint of governing, and even from the standpoint of our Constitution[104]." It looks like, after all the mistakes made, Donald Trump is trying to get out of the crisis and shift decisions and responsibilities to others as much as possible. He is no longer the wartime President. Does he feel like he lost the battle? The fact is that Donald Trump often turns his coronavirus press briefings into propaganda meetings. He does not hesitate to screen adulatory videos of himself to the journalists present, who often cannot believe what they are seeing. On April 13, *CNN* rebroadcasts these surreal images with an unambiguous caption: "Trump uses [coronavirus] task force briefing to try and rewrite history on coronavirus response." It looks like we are in Beijing…

On April 15, Trump found a scapegoat for his failures: WHO. He accuses the organisation of excessive delays and bias towards China and announces that day to suspend the financial contribution of the United States, which represents 22% of WHO's total assessed contributions for 2020–2021. He will confirm this decision on May 29 by declaring that his government "will be today terminating our relationship with the World Health Organization and redirecting those funds to other worldwide and deserving urgent global public health needs." No one is fooled: the President has found his scapegoat and wants the organisation to take responsibility for its own inconsistencies. An incredible decision, unacceptable in the midst of a global health crisis, which raises waves of protests. One protester among many was Bill Gates (on

---

[104]In the United States, which is a federation of 50 states, governors enjoy significant power, each state having its own criminal and civil law and managing, through its government, its internal affairs.

Twitter, @BillGates): "Halting funding for the World Health Organization during a world health crisis is as dangerous as it sounds. Their work is slowing the spread of Covid-19 and if that work is stopped no other organisation can replace them. The world needs WHO now more than ever." Richard Horton goes even further: according to him, Donald Trump is liable to a conviction for crimes against humanity for having decided to attack and weaken an organisation whose *raison d'être* is precisely to protect almost the entire world's population.

On April 17, discussions are going on in Washington about U.S. funding granted to the BSL-4 laboratory (biosafety level 4) of the Institute of Virology in Wuhan, apparently following a report in the Daily Mail.[105] BSL-4, the highest level of biosafety precautions, is applied to laboratories appropriate for work with pathogen organisms that may cause severe to fatal disease in humans. According to the article, a funding worth in total USD 3.7 million had been granted by the NIH to the EcoHealth alliance based in New York, which collaborates with researcher Shi Zhengli from the Wuhan Institute of Virology. The title of the research project is "Understanding the risk of bat coronavirus emergence" (grant number R01AI110964).[106] At the daily press conference on that day, Donald Trump launches his attack. "We watched this just an hour ago. It appears the decision was made by the previous President in 2015. We decided to stop the project immediately." The President did not grasp the fact that this research aims to investigate bat SARS-related coronaviruses. It cannot be more relevant in the current context as it might provide insights into the original source of the devastating outbreak. But in this political turmoil, there is no more room for basic science and collaboration with China…

In the context of the economic war that the United States is waging, all eyes turn to China, which has become a major political target. In Europe, Dominic Raab, who is acting Prime Minister as Boris Johnson convalesces, and Emmanuel Macron are following in Donald Trump's footsteps. The French President also criticises China's handling of the epidemic (while remaining very vague) and highlights the fact that no comparison is possible between countries where information flows freely and citizens can criticise their governments and those where the truth was suppressed: "Given these

---

[105] Owen G (2020, April 11) Wuhan lab was performing coronavirus experiments on bats from the caves where the disease is believed to have originated - with a £3 m grant from the U.S., The Daily Mail, https://www.dailymail.co.uk/news/article-8211257/Wuhan-lab-performing-experiments-bats-cor onavirus-caves.html?ito=social-facebook&fbclid=IwAR0nrOiS9AtOoaXHWbZNTLrlBl1FNx1_NHj dU1VN12kRG5qbQI1qlKNRj8E.

[106] https://taggs.hhs.gov/Detail/AwardDetail?arg_AwardNum=R01AI110964&arg_ProgOfficeCo de=104.

differences, the choices made and what China is today, which I respect, let's not be so naive as to say it's been much better at handling this. We don't know. There are clearly things that have happened that we don't know about," he says in an interview with the *Financial Times* published on April 16.[107]

We have clearly shown above that Chinese management has not been exemplary and that there are many grey areas. But, in this case, the French and American Presidents and the head of the British government should undoubtedly put their own houses in order... None of them can claim to have shone in their handling of the health crisis.

The federal capital, Washington, is in turmoil. Donald Trump looks destabilised. He announced on April 15 his intention to restart normal life, but a few days later he passes an executive order to suspend immigration. He claims to preside over the largest economy in the world's history, but one month later the unemployment rate more than doubles. While several states are still shut down, the justice department proclaims it would support legal action against governors who continue to impose strict social distancing rules. The President himself encourages the protests against physical distancing and other coronavirus stay-at-home measures that are developing all over the country. Donald Trump is obviously looking at his re-election in November, and he knows he will be judged on his economic record. He has hardly any word for the tens of thousands of his compatriots who are victims of Covid-19.

As a sign of the seriousness of the situation, the President no longer misses, from March 20 onwards, the almost daily briefings organised in the press room of the White House, where only a few accredited journalists are admitted, after a temperature check beforehand. These press points, which sometimes last more than two hours, feature a President in action against "an invisible enemy," surrounded by seasoned professionals, including Deborah Birx, a doctor and world-renowned Aids researcher who coordinates the coronavirus task force, and Anthony Fauci. Actually, it looks as if the President wants to project a positive image and restore his scientific reputation. Dr Fauci is renowned and respected; he is arguably the most listened-to person in the U.S. administration on the coronavirus. The immunologist is the scientific face of America's Covid-19 response: his positions are clear and robust. On the hydroxychloroquine issue, Fauci says that "scientifically speaking," he does not see how we can demonstrate "any benefit from this treatment." But his position apparently did not deter the President, who publicly admits on May 19 to taking "a pill every day" since the beginning of the month.

---

[107]Mallet V and Khalaf R (2020, April 16) Emmanuel Macron says it is time to think the unthinkable, Financial Times, https://www.ft.com/content/3ea8d790-7fd1-11ea-8fdb-7ec06edeef84.

The immunology expert, a member of the White House coronavirus task force, has made no secret of the difficulty of working with the Trump administration. In an interview with Science magazine,[108] he explains the basic—almost educational—work he has to do when standing with the President in the press room. The interview sheds an interesting light on how scientists work with politicians. "I do not disagree [with him] in the substance," Anthony Fauci explains, diplomatically, to journalist Jon Cohen. "It is expressed in a way that I would not express it, because it could lead to some misunderstanding about what the facts are about a given subject [for example, when the President calls SARS-CoV-2 the 'Chinese virus'—author's note]. Personally, I will never say that it is a Chinese virus and journalists know it well." Then Jon Cohen asks Fauci how he manages false statements put forward by the President, for example when he says China should have told us 3 to 4 months earlier and that they were 'very secretive'." "I know, Fauci tells the journalist, but what do you want me to do? I mean, seriously Jon, let's get real, what do you want me to do? The way it happened is that after he made that statement, I told the appropriate people, it doesn't comport, because 2 or 3 months earlier would have been September. The next time they sit down with him and talk about what he's going to say, they will say, "By the way, Mr. President, be careful about this and don't say that." But I can't jump in front of the microphone and push him down. OK, he said it. Let's try and get it corrected for the next time."

Then Jon Cohen asks questions about how they are preparing the press briefings and are working in the task force. Anthony Fauci explains: "We meet every morning and work for about an hour and a half. We sit down for an hour and a half, go over all the issues on the agenda. And then we proceed from there to an anteroom right in front of the Oval Office to talk about what are going to be the messages, what are the kind of things we're going to want to emphasise? Then we go in to see the President, we present [our consensus] to him and somebody writes a speech. Then he gets up and ad libs on his speech. And then we're up there to try and answer questions." The reporter then reminds Anthony Fauci that at the press conference the week before, he put his hands over his face when Donald Trump referred to the "Deep State Department" (a popular conspiracy theory in the United States—and valued by Donald Trump—that pretends that the real power is held by a few officials in agencies and businesses, a sort of hidden government). The

---

[108]Cohen J (2020, March 22) 'I'm going to keep pushing.' Anthony Fauci tries to make the White House listen to facts of the pandemic, Science, https://www.sciencemag.org/news/2020/03/i-m-going-keep-pushing-anthony-fauci-tries-make-white-house-listen-facts-pandemic.

episode has become a buzz (viral?) on the Internet. Jon Cohen: "Have you been criticised for what you did?" "No comment," Anthony Fauci replies.

The members of the task force are regularly confronted with a difficult dilemma: how to be respectful of their President (and therefore not to contradict him publicly) without correcting inaccurate or even dangerous information. But what happened on April 23 will no doubt remain as a textbook case. That day, during the daily conference, Donald Trump surprises everyone, with an intervention which he keeps totally secret, suggesting that Covid-19 might be treated by injecting into the body a disinfectant such as bleach or isopropyl alcohol[109]: "I see [on the presentation projected on the screen—author's note] the disinfectant where it knocks out [the coronavirus] in a minute. One minute! And is there a way to do something like that, by injection inside or almost a cleaning? So, it'd be interesting to check that," he tells the press. Donald Trump also proposes to hit the body with a tremendous powerful light like ultraviolet, "supposing you brought the light inside of the body, which you can do either through the skin or in some other way. And [turning to Deborah Birx] I think you said you're going to test that too. Sounds interesting," the President continues. Then, Donald Trump turns again to Dr Birx and asks if she has ever heard of using "the heat and the light" to treat the coronavirus. She responds briefly (before the President cuts her off): "Not as a treatment, but fever is a good thing …". However, the report distributed by the White House to the press in the evening credits Deborah Birx with the following quote: "This is a cure." The next day, the report is sent back to the press with Dr Birx's correct answer: "Not as a treatment". After this episode, Donald Trump will cancel his press briefing for three days, to resume on April 27. That day, when asked by a reporter about the reason for a surge in phone calls to the poison control centres in Maryland after the [April 23] press conference, the President says he does not know. The next day, April 28, Vice-President Mike Pence, who heads the task force, draws criticism from the press by visiting the prestigious Mayo Clinic in Minnesota without wearing a mask, despite the medical facility's policy requiring visitors to wear face coverings.

On April 24, *CNN* quotes a member of the coronavirus task force who, on condition of anonymity, says that they are often caught off guard by the President's comments, which they sometimes find "surreal." The experts recognise that they are unable to predict what Donald Trump will absorb from their briefings and discussions as he often picks an anecdotal detail that he will put forward, in his own way, to the televised press conference a few minutes

[109]BBC News (2020, April 24) Coronavirus: Outcry after Trump suggests injecting disinfectant as treatment, https://www.bbc.com/news/world-us-canada-52407177.

later. They are regularly faced with the choice of either publicly contradicting the President or letting inaccurate and even dangerous information go uncorrected. The consequences of either seem unpalatable. Thus, speaking at the Fox News channel's (virtual) Town Hall event on May 3, the President announces that a vaccine against the coronavirus would be available at the end of the year (which, by the way, will turn out to be correct).[110] But he also adds, "Now, the doctors would say, 'Well, you shouldn't say that.' I'll say what I think." The fact is that Donald Trump is desperately hoping that the promise of a future vaccine will help Americans to accept the heavy toll that is to come as confidential administration documents forecast a record-high of three thousand deaths per day in May.

On May 4, thunder claps in the White House: Anthony Fauci declares in an interview with National Geographic that "the best evidence shows the virus behind the pandemic was not made in a lab in China," therefore openly contradicting Donald Trump and his Secretary of State Mike Pompeo[111]: "Everything about the stepwise evolution over time strongly indicates that [this virus] evolved in nature and then jumped species," the immunologist says. He does not believe that someone found the coronavirus in the wild, brought it to a lab, and then it accidentally escaped. As, when talking about SARS-CoV-2, the President has crossed out the word "Corona" and replaced it with "Chinese," one cannot possibly ask for a better demonstration, absurdly, of the lack of consideration and attention that Donald Trump shows to his key advisor on the pandemic.

At the start of May, the President has his eyes fixed on the economy. He cannot hide his eagerness to reopen businesses. He has already lost ground in the field of health (80 000 deaths as of May 12, 2020, whereas he hoped to stay below the 100 000 mark a month earlier); if he also falls short on the economy, he knows his chances of being re-elected will be near zero. The President then ventures down a dangerous path: defending the idea that "the cure cannot be worse than the problem," he seems to concur with the idea of sacrificing a number of lives to allow the majority to return to work. Incidentally, he announces the gradual reduction in the activities of the coronavirus task force. No one is fooled: Donald Trump wants to disband the group of experts after the public declarations of Anthony Fauci on the origin

---

[110] https://www.whitehouse.gov/briefings-statements/remarks-President-trump-fox-news-virtual-town-hall/.

[111] Akpan N and Jaggard V (2020, May 4) No scientific evidence the coronavirus was made in a Chinese lab, National Geographic, https://www.nationalgeographic.com/science/2020/05/anthony-fauci-no-scientific-evidence-the-coronavirus-was-made-in-a-chinese-lab-cvd/.

of the virus. However, this announcement causes a public outcry, forcing the President to change his mind...

On May 12, during a virtual hearing before the U.S. Senate Committee on Health, Education, Labor and Pensions, Anthony Fauci warns against a too hasty restart of the economy in the midst of a pandemic: "The consequences could be really serious," the expert says. He then explains that the coronavirus death toll is likely higher than the official 80 000 deaths. The expert therefore contradicts Donald Trump who is claiming that the numbers of cases are falling everywhere, allowing the country to reopen and the economy to restart "safely."

Faced with the enemy Covid-19, *Private Trump* practically disintegrated within a matter of weeks. His explosive cocktail of an oversized ego and contempt for the elites, arguably coupled with a deep inferiority complex, turned out to be a more toxic poison than the virus, making him deaf to medical and scientific advice. The U.S. President has only succeeded in convincing the whole world of his utter incompetence and obsession with chasing an illusory leadership in vain. The unwavering support shown to him by the conservative electorate is breaking down at the pace of the pandemic and is highlighting the deep divisions of American society. The crisis also brings the country's weaknesses in the spotlight in several important fields such as diplomacy, economy, social, health care. And science too.

Nevertheless, the United States is working twice as hard to win the vaccine race. On April 29, Donald Trump launches the "Operation Warp Speed", endowed with ten billion dollars, which aims to develop a vaccine in just a few months—but for "America first." The idea dates back to February 15, 2020, when the President's trade adviser, Peter Navarro, proposed a "Manhattan vaccine project" to boost the U.S. research effort in this area[112]. At that time, the officials did not have a precise idea about which vaccine would be the frontrunner, but expected to select a dozen "good candidates" from current research. They do not rule out working with foreign companies—apart from the Chinese "of course." Unlike the European Union and WHO which promote international work, the promoters of Warp Speed aim to ensure that American people are first in line to benefit from Covid-19 vaccines. A regrettable reflex not only from a humanitarian point of view, because it risks not creating the conditions for a universal vaccine accessible to all, but also from a scientific point of view because it may not then

---

[112]Manhattan is the code name of the research project led by the United States with the support of the United Kingdom and Canada that produced the first atomic bomb during World War II. Launched in 1939, the project mobilised up to one hundred thousand people and cost about USD two billion, or around thirty billion dollars in today's values.

benefit from innovations and research developed elsewhere in the world. At the same time, the French company Sanofi announces that the first sales of its hypothetical vaccine will be for the United States, "because they took a risk to finance this research before the others, from the month of February," Paul Hudson, the company's managing director, tells Bloomberg. A statement that raises public outcry in France. Because the stakes are obviously colossally high, and not just financially, many states are showing "vaccine nationalism."

From this point of view, the United States clearly chose to go its own way. But they are not the only ones—far from it. Fortunately, there are also a couple of multilateral initiatives. At the beginning of July, 2020, the European Commission announces it had started, on behalf of the twenty-seven Member States, negotiations with the American pharmaceutical company Gilead Sciences to buy 30 000 doses of the drug remdesivir, the first antiviral to have shown positive results—although actually very modest—on patients with severe forms of Covid-19. These discussions follow the announcement by the United States of the pre-purchase of almost all of the world production for the period from July to September 2020. In August, the Commission reaches agreements with five pharmaceutical companies working on a vaccine candidate to ensure that all EU states can be supplied quickly and in sufficient quantities as soon as a vaccine is deemed effective. For example, the Commission signs a first contract with the German company CureVac for the purchase of 400 million doses of vaccine. By following a single and non-commercial approach, in respect of usual procedures, the European executive is acting in the interests of the European Union and even beyond. But it is clear that, in such a tense situation, disadvantaged countries are hardly able to make their voices heard and ensure that their needs are taken into account. Along with Gavi (the Vaccine Alliance) and WHO, the EU is also a major contributor to COVAX, a global collaboration which aims to secure 1.3 billion doses of vaccine for 92 low and middle-income countries by the end of 2021.

At the height of the crisis, Donald Trump likes to repeatedly point out that no one could have predicted such a catastrophe. However, this is not true. Peter Navarro, his top trade adviser, warned Trump administration officials in late January that the coronavirus crisis could cost the United States trillions of dollars (millions of billions) and put millions of Americans at risk of illness or death.[113] Of course, it is always easy, after an event, to find people who understood or predicted it before its time. But Peter Navarro's memo dates

---

[113]Haberman M (2020, April 6) Trade Adviser Warned White House in January of Risks of a Pandemic, The New York Times, https://www.nytimes.com/2020/04/06/us/politics/navarro-warning-trump-coronavirus.html?referringSource=articleShare.

from January 29, 2020, which is just eight days after the first U.S. case was identified and less than a week after Hubei was placed in lockdown. At that time the epidemic was just starting to spread into Europe. The Trump administration was also beginning to work on measures the United States could take to deal with the looming pandemic. Navarro's note therefore did not fall *ex abrupto* in a context of "business as usual" but at a time when worrying signs were coming from several countries and from several sources around the world. This was apparently the U.S. administration highest level of alert, made by one of the President's own advisers. Governing is also knowing how to surround by the right people and listening to them…

Specifically, Peter Navarro details his concerns in two memos, one on January 29—the very same day that Trump set up the coronavirus task force—and another on February 23. The first memo explains that "the lack of immune protection or an existing cure or vaccine would leave Americans defenceless in the case of a full-blown coronavirus outbreak on U.S. soil. […] This lack of protection elevates the risk of the coronavirus evolving into a full-blown pandemic, imperilling the lives of millions of Americans," Navarro writes. The adviser ends his note by asking the administration to decide to what extent it wishes to take "aggressive" measures, considering that it may be possible to contain the epidemic as is done for seasonal flu, with limited human and economic cost. However, he concludes with a very clear warning that the risk of a "worst-case scenario" should not be ruled out for the coming pandemic, with the virus killing more than "half a million" Americans. Navarro is unfortunately quite right…

On January 28, as Navarro is writing his note, Carter Mecher, a Senior Medical Adviser for the Office of Public Health, sends an email to several dozen of his colleagues, all public health experts working for the government or for American universities: "The projections of the epidemic already defy understanding." Mecher writes just a week after the first case was confirmed in the United States, but his conclusions are unambiguous and he urges top officials in the public health administration to consider drastic action.

Numerous articles confirm it: as early as January 2020, while Donald Trump minimises the epidemic risk and focuses on other subjects, several members of his administration—advisers to the White House and experts in cabinets and in intelligence agencies—have identified the threat, sent alerts and concluded that there was a need for rapid and far-reaching action.[114] The National Security Council also sent notes in mid-January on the arrival

---

[114]Lipton E, Sanger DE, Haberman M, Shear MD, Mazzetti M and Barnes JE (2020, April 11) He Could Have Seen What Was Coming: Behind Trump's Failure on the Virus, The New York Times, https://www.nytimes.com/2020/04/11/us/politics/coronavirus-trump-response.html.

of the epidemic and raised the option of lockdown in large cities. Then came the memos from Peter Navarro. Then the Secretary of Health and Human Services, Alex Azar, discussed the risks of Covid-19 directly with the President by telephone on January 30. But Donald Trump dismisses his interlocutor for being "too alarmist." Alex Azar announced in February that a network to monitor the epidemic in major cities would be set up, but only after a delay of several weeks.

A few months later, the Presidential line will be shattered when, on September 9, 2020 *CNN* publishes excerpts from journalist Bob Woodward's new book "Rage," devoted to Donald Trump's presidency.[115] These few pages show the President was well aware that the coronavirus is dangerous, highly contagious and "more deadly than even your strenuous flus." So, the President repeatedly lied to his fellow citizens and deliberately delayed taking actions he knew would have a big impact on the economy. In his own defence, Donald Trump responded by saying that it was necessary to show leadership and create trust. "To be honest with you, I wanted to always play it down," Trump told Bob Woodward on March 19, 2020, although he had declared a national emergency over the virus days earlier. "I still like playing it down, because I don't want to create a panic." Trump's admissions are therefore in stark contrast to his frequent public comments at the time insisting that the virus was "going to disappear" and "all work out fine."

However, early March 2020, Donald Trump realises that the epidemic will soon explode. Yet sources close to the government explain that in internal discussions, economic and political considerations still prevail over public health. Even when the number of cases started to rise sharply, such as in February, it will take several weeks for the President's advisers to convince him that without swift action the toll will be much higher. Internal rivalries, the President's general mistrust of experts, his desire to preserve relations with China at that time and confidence in his own judgment are all the factors that contributed to slowing down the decision at the highest level of the state.

There is therefore no doubt that the American President was well informed of the situation and the possible risks by mid-January, and even before. It cannot be otherwise for the most developed nation in the world.

On May 7, 2020, Rick Bright, a vaccine and infectious disease specialist, accuses the administration of having ignored the warnings he issued in January about the dangers of the coronavirus. A former director of the Biomedical Advanced Research and Development Authority (BARDA),

---

[115]Gangel J, Herb, J and Stuart E (2020, September 9) 'Play it down': Trump admits to concealing the true threat of coronavirus in new Woodward book, CNN politics, https://amp.cnn.com/cnn/2020/09/09/politics/bob-woodward-rage-book-trump-coronavirus/index.html?__twitter_impression=true.

which belongs to the U.S. Department of Health and Human Services, Rick Bright was dismissed from his post on April 20, due, according to the doctor, to his reluctance to promote treatment with hydroxychloroquine, which is favoured by the President. On May 15, before the House of Representatives' sub-committee on health, Rick Bright gives his own explanations about his abrupt change of professional responsibilities: "I believe this transfer was in response to my insistence that the government invest the billions of dollars allocated by Congress to address the Covid-19 pandemic into safe and scientifically vetted solutions, and not in drugs, vaccines and other technologies that lack scientific merit," he says in his statement. Then, Dr Bright accuses his superiors of using public money for political purposes or cronyism. Funding worth USD 20 million was reportedly granted to a company located in Florida, despite having been rejected by Rick Bright as he demonstrated that the work had no real scientific value. He tells the Representatives that he was pressured to direct money toward hydroxychloroquine, one of several "potentially dangerous drugs promoted by those with political connections" and repeatedly describes by the President as a potential "game changer" in the fight against the virus. "I am speaking out because to combat this deadly virus, science—not politics or cronyism—has to lead the way (Fig. 3.4)."

Fig. 3.4 Donald Trump, Deborah Birx (left) and Anthony Fauci (right) at the White House coronavirus task force briefing on Friday May 15, 2020 (© Stefani Reynolds/CNP via ZUMA Wire)

On June 8, 2020, as most of the states have completely lifted restrictions, New York City celebrates the end of shutdown and confirms the gradual restart of the economy. However, the outbreak is still far from being suppressed but the President seems to have turned the page and has ceased government communication on the pandemic since the end of April. On June 20 in Tulsa, Oklahoma, Donald Trump marks the resumption of his political meetings for the November elections. Time is running out: a poll released on June 24 shows that only 37% of Americans have a favourable opinion of Donald Trump. The President is racking up criticism, including within his own ranks, for his handling of the crisis and his attitude following the death of George Floyd, a young African-American man, who died on May 25 after being handcuffed and pinned to the ground during a police check in Minneapolis, Minnesota. Most notably, Joe Biden, the Democratic Presidential candidate is 14 points ahead in voting intentions. In Tulsa, the President no longer draws the crowds. And some members of the Republican Party make no secret of their feelings about Donald Trump, seeing the failure to meet popular expectations as a demonstration of the incompetence of the President's team. By instrumentalising the disease for political ends, the President has exacerbated social tensions, scrambled health messages and disrupted coordination between states.

On June 20, the epidemic grows exponentially in 36 states, especially those that have reopened fast. Technically speaking, this is not a "second wave" since the first one has not yet ended… Several states ask their inhabitants to stay at home and make it compulsory to wear a mask in public, which has been recommended by health authorities for weeks. But the beaches of Florida remain open and the President still refuses to wear the mask, and still urges his supporters to do the same, as we saw during the Tulsa meeting.

Always true to self, Donald Trump has once again attributed this surge in new infections to the increased access to testing and to the fact that the tests are "much better" than in any other country. However, on July 11, the President doubles back, appearing for the first time in public with a protective mask, a strong symbolic gesture at a time when the United States is still the country most affected by the pandemic, with more than three million infections and one hundred and thirty thousand deaths. A failure that is partly due to disparate political decisions across the country and states and also to the fact that a percentage of the population is reluctant to follow public health guidance. Even wearing a face mask has become the subject of partisan debate.

During this period, Anthony Fauci often contradicts the President, no longer hesitating to publicly declare that the situation in the country is very serious. His appearances at the White House become few and far between. Especially since the President's entourage tried to torpedo the reputation of the scientist: on 12 July, the press service distributes an "anonymous" list of errors of assessment that the infectious disease specialist allegedly made since the start of the epidemic. In a July 14 op-ed published in USA Today, Peter Navarro sharply criticises Dr Fauci, saying that the scientist has been "wrong about everything" and that he is listening to the doctor's advice "only with scepticism and caution."[116] In reality, Trump and Fauci cannot do without each other. Removing Anthony Fauci from the White House task force is in theory possible but it would be a strategic error for which Donald Trump does not want to take responsibility. And Fauci must remain close to the President to remind the population as much as possible of prevention and protection measures and thus help the Americans to fight the epidemic. The increased distance between the two men also reflects the fact that the hot spot of the crisis is now at the level of the governors since, in the absence of clear guidelines from the President, they have the last word.

The fact remains that the President's entourage no longer refrains from considering Anthony Fauci as useless or even incompetent. On August 2, House of Representatives Speaker Nancy Pelosi accuses Deborah Birx and the task force of being responsible for the White House's misinformation about the coronavirus, implying that she and Fauci are not weighing in against the President. To which Doctor Birx replies on *CNN* that Covid-19 is spreading widely in the country, including in rural communities. She calls on Americans to take more precautions, especially advising infected people to wear a mask at home if they live with vulnerable relatives. Taking this recommendation as a criticism, Donald Trump immediately accuses Deborah Birx on Twitter of "having taken the bait," in other words of having fallen into the trap set by the Democrats. The distance between politics and science is increasingly blatant. And this also has repercussions at the level of doctors and hospitals, if only by the lack of means, tests and materials. Even the famous CDC is unable to fulfil their mission. Since the beginning of the year, the US government is attempting to cut their budget, which has sparked a volley of green wood from the scientific community, as the normal functioning of CDC is indeed essential to prevention and control of the epidemic.

---

[116]Navarro P (2020, July 14) Anthony Fauci has been wrong about everything I have interacted with him on: Peter Navarro, USA Today, https://eu.usatoday.com/story/opinion/todaysdebate/2020/07/14/anthony-fauci-wrong-with-me-peter-navarro-editorials-debates/5439374002/.

In early July, the National Academy of Sciences updates an open letter, signed by 1 220 scientists, who accuse the Trump administration of "denigration of scientific expertise and harassment of scientists."[117] These are very harsh words from an organisation that used to keep politics at arm's length. The signatories call on the federal government to "restore a science-based policy," adding that the "dismissal of scientific evidence in policy formulation has affected wide areas of the social, biological, environmental and physical sciences." This open letter was originally a response from scientists to Donald Trump's decision to withdraw from the Paris climate agreement. But scientists' disappointment with the U.S. government's response to the epidemic led its authors to update the letter and invite colleagues to sign it. This position is necessary, but not sufficient. The scientific community should work actively, at international level, on the conditions which should be put in place to avoid this happening again. We will come back to this in the last chapter.

Still, the situation is not improving at all as the country is approaching a hundred thousand confirmed cases on a daily basis at that time. In Texas, Houston, a city which has a medical capacity higher than New York or Washington, is totally overwhelmed. The good news is that a significant proportion of the population is listening to experts and do follow their recommendations, especially when it comes to wearing a mask. Thus, according to a survey carried out by the University of Quinnipiac and published on July 15, 65% of Americans say they trust the information Anthony Fauci is providing about the coronavirus. Conversely, the same proportion admits not trusting the information given by President Trump.[118]

On July 15, the Trump administration instructs hospitals to reroute their coronavirus data first to the Trump administration (the Department of Health and Human Services, HHS) instead of sending it directly to the CDC. Officially, the change aims to streamline data gathering and better assist the White House coronavirus task force. Nobody is fooled: the decision confirms the politicisation of the crisis and risks leading to a loss of transparency in the health data that will be provided by the United States.

The day after, Donald Trump's fourth spokesperson, Kayleigh McEnany explains in the press room that the President is very keen on seeing schools reopen and that "science should not stand in the way of that."[119] McEnany is then citing a scientific study published in JAMA Paediatrics which she claims

---

[117] Statement to Restore Science-Based Policy in Government, https://scientistsforsciencebasedpolicy.org/.

[118] Quinnipiac University Poll (2020, July 15) Biden Widens Lead Over Trump To 15 Points In Presidential Race, https://poll.qu.edu/national/release-detail?ReleaseID=3666.

[119] Press Briefing by Press Secretary Kayleigh McEnany (2020, July, 16) https://www.whitehouse.gov/briefings-statements/press-briefing-press-secretary-kayleigh-mcenany-7-16-2020/.

is supporting the reopening of schools: "The science is on our side here, she says, and we encourage localities and states to just simply follow the science, open our schools." However, she misses the point here. The article actually shows that the risk for children to get a critical illness is far less from Covid-19 than from seasonal flu, and most experts worry about returning to classrooms not because of the risk to the students themselves, but because of the risk that children could worsen the spread of coronavirus by bringing it home and passing it to older relatives.

And on July 19, a surreal television sequence which now appears as one of the pearls of the pandemic: "I think we have one of the lowest mortality rates in the world," declares Donald Trump on the television channel American Fox News,[120] known for its unwavering support for the U.S. President. "This is not true sir," journalist Chris Wallace replies. "We have the seventh highest mortality rate in the world; [...] we have 900 deaths in a single day," he adds. "We will check," says Donald Trump. He then asks Kayleigh McEnany to give him the mortality figures. He receives a sheet of paper and concludes: "There you are. Number one low mortality...". The interview makes a buzz on social networks, at a time when the country has more than 140 000 Covid-19 deaths. I believe that Donald Trump knew that his country was number one in terms of Covid-19 mortality (actually the crude number of deaths) but he thought—and probably sincerely—that first position means a better score... And yet another twist on July 21, when Donald Trump once again turns on his heels during the first press briefing devoted to the pandemic since the end of April, finally recognising, after weeks of denials, that the health crisis "will probably unfortunately get worse before it gets better. Something I don't like saying about things but that's the way it is."

Only three months away from the Presidential elections, Donald Trump is trying at all costs to restore his image and put pressure on the development of vaccines and treatments. Thus, on Sunday August 23, 2020, during a press conference described as "very unusual" by *Science magazine*, Donald Trump announces a so-called Emergency Use Authorisation (EUA) for blood plasma transfusion from people who recovered from the coronavirus to hospitalised patients. This treatment is already widely used in the United States and other countries, but its actual effectiveness is still debated. The goal is to transfuse the antibodies contained in the convalescent plasma to allow patients to eliminate the virus more quickly and to limit the effects on the body. While the treatment has already produced results, its exact effectiveness is questioned by the scientific and medical community and by WHO

---

[120] https://video.foxnews.com/v/6172996717001#sp=show-clips.

because it presents risks of side effects and transmission of infectious agents. Still, the FDA decides to issue an emergency clearance, despite the lack of convincing results from the major clinical trials devoted to this therapeutic strategy. Quite bizarrely, the FDA announcement is released a few minutes before the President's press conference. Is it to remind us that the EUA is the agency's responsibility and not that of the Head of State and to avoid giving the impression that the agency was under pressure from the White House? FDA Commissioner Stephen Hahn dismisses any political pressure at the press conference. However, in its press release, the FDA cautiously specifies that "it is reasonable to believe that Covid-19 convalescent plasma may be effective in lessening the severity or shortening the length of Covid-19 illness in some hospitalised patients" while recommending that the ongoing randomised clinical trials must be continued for "the definitive demonstration of safety and efficacy of Covid-19 convalescent plasma."[121] This does not prevent the President, speaking at the daily press briefing, to claim victory: "I'm pleased to make a truly historic announcement in our battle against the China virus that will save countless lives."

The next day, August 24, another controversy arises about SARS-CoV-2 testing, as the CDC publishes on their website updated guidelines that now state that a patient who has been in close contact ("within 6 feet") of a person with a Covid-19 infection for at least 15 min but does not show symptoms, that person "does not need to be tested" unless they are vulnerable or their health care provider or state or local public health officials recommend it. Until then, the CDC's position was that people without symptoms who were exposed to the virus should be "quickly identified and tested." Clearly a change from the current screening strategies, which aim to test asymptomatic people which can represent nearly half of confirmed cases. On August 26, CDC Director Robert Redfield tries to clarify these new recommendations by issuing a statement in which he says that "Everyone who needs a Covid-19 test, can get a test" and also that "Everyone who wants a test does not necessarily need a test." Apparently, Robert Redfield tells some news outlets that "testing may be considered for all close contacts of confirmed or probable Covid-19 patients." However, on August 27, the CDC's website had still not been corrected. Administration officials explain that "not necessarily" is consistent with "may be considered." Dr Tom Frieden, former director of the CDC, is furious: "The CDC guidance is indefensible," he writes on Twitter

---

[121] US Food and Drug Administration (2020, August 23) FDA Issues Emergency Use Authorization for Convalescent Plasma as Potential Promising COVID–19 Treatment, Another Achievement in Administration's Fight Against Pandemic, https://www.fda.gov/news-events/press-announcements/fda-issues-emergency-use-authorization-convalescent-plasma-potential-promising-Covid-19-treatment.

on August 27. "No matter who wrote it and got it posted on the CDC site, it needs to be changed." Anyway, the new recommendations go wrong, as a majority of states decide not to implement them. A *New York Times* article on September 17 confirms what everyone was thinking: it is the HHS (i.e., the Trump administration) that imposed the controversial directive on the U.S.' leading public health authority as the new guidance was not written by the CDC's scientists and has been posted despite their objections.[122]

There is no doubt that, in this context, strong pressure was applied to the companies and the relevant administrations to ensure that a vaccine could be successfully announced before November 3, the date of the Presidential elections. The best demonstration is that, on September 8, the CEOs of nine pharmaceutical groups (AstraZeneca, BioNTech, GlaxoSmithKline, Johnson & Johnson, Merck, Moderna, Novavax, Pfizer and Sanofi) publish a joint pledge that they oppose any rushed authorisation of a vaccine, anticipating political pressure. The bosses say in particular that they commit themselves "to uphold the integrity of the scientific process as they work towards potential global regulatory filings and approvals of the first Covid-19 vaccines." They also specify that they will "only submit for approval or emergency use authorisation after demonstrating safety and efficacy through a Phase 3 clinical study that is designed and conducted to meet requirements of expert regulatory authorities such as FDA." Clearly, the big pharmaceutical groups are worried about this politicisation of science in the United States. On the same day, the British laboratory AstraZeneca announces it is suspending the phase 3 clinical trial of its vaccine developed with the University of Oxford, due to the negative reaction of one of the patients in the United Kingdom.

On October 2, 2020, the international media goes wild: Donald Trump and his wife Melania had tested positive for Covid-19 the night before. After Boris Johnson and Jair Bolsonaro, Donald Trump is now the third Head of State or Government who downplayed Covid-19 and then contracted it. Several close collaborators of the President are also positive. At the end of the day, the President is admitted to Walter Reed Military Hospital in Bethesda, a suburb of Washington. According to Jim Acosta, *CNN* correspondent at the White House, Donald Trump is tired and has some breathing difficulties. Kayleigh McEnany, who also tested positive, says that doctors have prescribed the antiviral remdesivir as well as dexamethasone, a corticosteroid effective against severe forms of Covid-19. Jon Cohen claims on Twitter that

---

[122]Mandavilli A (2020, September 17) C.D.C. Testing Guidance Was Published Against Scientists' Objections, The New York Times, https://www.nytimes.com/2020/09/17/health/coronavirus-testing-cdc.html.

the President also allegedly received an injection of eight grams of mono-clonal antibodies produced and tested by the company Regeneron, which suggests that his state of health may be worrying. Just one month before the Presidential election, Covid-19 is entering the campaign directly, with the probable cancellation of Donald Trump's next political meetings. The Demo-cratic candidate, Joe Biden, receives a negative test three days after the first Presidential debate which took place on 30 September in Cleveland, Ohio. Joe Biden tells the press that he is going to continue his campaign, after wishing his opponent a speedy recovery and recalls in passing the impor-tance of wearing a mask. A few days earlier, Donald Trump announced the imminent end of the Covid-19 outbreak...

The President leaves the hospital on October 5 and immediately addresses the American people, on Twitter as usual: "Do not be afraid of Covid. Don't let it dominate your lives." In any case, the episode gives the impression that the White House is like a golden palace and has still not taken the measure of the epidemic. Donald Trump's return to business also marks the return of controversy. The fact that he took off his mask on the doorstep of the White House that evening shocked many people. Later, the President affirms once again on Twitter, in total conflict with the facts, that the flu kills more than Covid-19. For his part, Joe Biden recalls that "wearing a mask is not a political statement. It's a scientific recommendation. Let's end the politics and follow the science. Social distancing isn't a political statement. It's a scientific recom-mendation." The coronavirus will clearly be at the heart of the Presidential campaign. And, fearing that the announcement of the long-awaited vaccine will be postponed until after November 3, the White House is blocking new and stricter guidelines proposed by the FDA for the emergency release of the future vaccine. For its part, the CDC decides to maintain the usual standards and only authorise a vaccine if trials show that it reduces the incidence of Covid-19 by at least 50%.

At the same time, a declaration is signed on October 4 by tens of thou-sands of American scientists and health professionals to relaunch the debate on herd immunity. The authors plead for the lifting of lockdown restric-tions on young and healthy people while focusing protection measures on the elderly. The aim is to propose a (new) strategy to combat Covid-19. It so happens that one of the signatories of what is known as the "Great Barrington Declaration,"[123] named after the Massachusetts' city where the signing took place, is Dr Scott Atlas, the White House science adviser. The main thesis, i.e., allowing non-vulnerable people to resume active life in order to develop

---

[123] https://gbdeclaration.org/.

herd immunity, is indeed in line with the ideas defended by the President. No wonder that this initiative was well received by the Trump administration. But this approach is hardly unanimous among experts: on October 14, a hundred of scientists sign the "John Snow Memorandum" (there are now more than 7 000 signatories), in which they qualify the declaration as "a dangerous fallacy unsupported by scientific evidence."[124] Coincidence or not, on October 19, the President violently criticises Anthony Fauci during a telephone exchange with his campaign staff from a hotel in Las Vegas and to which several American journalists had access: "People are tired of hearing Fauci and all these idiots." The President adds: "Fauci is a disaster who has been around for 500 years. If I listened to him, we'd have 500,000 deaths."[125] Of course, Donald Trump knows that it would be counterproductive to dismiss Anthony Fauci within two weeks of the Presidential election. Ironically, at almost the same time the President was fuming on the phone, Dr Fauci is awarded the National Academy of Medicine's first-ever Presidential Citation for Exemplary Leadership, during a virtual ceremony.

A few weeks before the end of his mandate, Donald Trump is making a speech that is almost the exact opposite of the reality and the scientific evidence. He declares, in front of mostly unmasked crowds, that "the disease is already disappearing" and no longer threatens the health of Americans and the nation's economy. This is what he says while the number of cases is booming again in the country, the number of deaths is also starting to rise, the stock market has collapsed and some of the team of his Vice-President, Mike Pence, have been infected. However, for the President, the United States is "rounding the turn," a message that he repeated with shocking consistency throughout 2020. On November 30, Scott Atlas resigns from his post in the Trump administration. Dr Atlas was widely challenged in the scientific and medical community and frequently clashed with top government scientists, including Anthony Fauci and Deborah Birx, in particular for having attacked public health measures such as masks, stay-at-home orders and social distancing. He called on residents of Michigan to "rise up" against restrictions put in place by Governor Gretchen Whitmer, who had been the target of a kidnapping plot, leading to calls for his firing.

On November 3, 2020, the Presidential elections takes place in a very tense climate and in a country more divided than ever. Four days later, when the

---

[124] https://www.thelancet.com/journals/lancet/article/PIIS0140-6736(20)32153-X/fulltext#sec1.

[125] After this phone call, Donald Trump even said that the U.S. would have "700 000 or 800 000 deaths" if he had listened Dr Fauci: Collins K and Liptak K (2020, October 20) Trump trashes Fauci and makes baseless coronavirus claims in campaign call, CNN, https://edition.cnn.com/2020/10/19/politics/donald-trump-anthony-fauci-coronavirus/index.html.

vote count is not yet complete in five states, Joe Biden is declared by the media as the 46th President of the United States, having won that day 279 Electoral College votes out of the 538 allotted to the 50 states and the District of Columbia, that is to say nine more than the absolute majority (270). It was a huge relief, especially in the scientific and medical community, because Joe Biden had clearly made Covid-19 (along with global warming) a priority on his program. Prestigious scientific journals like *Nature* and *Scientific American* had clearly showed their support for the Democratic candidate. Without delay, the day after their victory was declared in the U.S. election, future President Joe Biden and Vice-President Kamala Harris announce a Covid-19 advisory board co-chaired by Vivek Murthy, former Surgeon General, David Kessler, former Commissioner of the FDA and Marcella Nunez-Smith of Yale University. The board consists of thirteen members, scientists and doctors, including Rick Bright, the expert who had resigned from the administration a month before. While the United States is experiencing an exponential growth in the epidemic, with a record of more than 120 000 positive cases and 61 000 hospitalisations in 24 h, the President-elect wants an action plan to come into force on January 20, 2021, the day of his Presidential inauguration.

And, for the first time, clear recommendations are sent from the top to the U.S. population: "Please, I implore you, wear a mask," Joe Biden said on November 9, 2020. "Do it for yourself. Do it for your neighbour. A mask is not a political statement, but it is a good way to start pulling the country together. We can save tens of thousands of lives if everyone would just wear a mask for the next few months. Not Republican or Democratic lives, American lives […]. It's time to end the politicisation of basic, responsible public health steps like mask-wearing and social distancing." The same day, Donald Trump reacts to the announcement of Pfizer and BioNTech concerning a vaccine "effective at 90%" and accuses the leaders of the two companies, plus the FDA and the Democratic Party, of having "conspired" to deliberately release the news after Election Day. In *Science*, Jon Cohen shows that there is no evidence supporting Donald Trump's claim.[126] Whether it is a coincidence or a consequence of political change, the CDC acknowledges on November 10 that face masks also protect those who wear it by filtering down inhalation of airborne droplets.[127]

---

[126]Cohen J (2020, November 11) Fact check: No evidence supports Trump's claim that Covid-19 vaccine result was suppressed to sway election, Science, https://www.sciencemag.org/news/2020/11/fact-check-no-evidence-supports-trump-s-claim-Covid-19-vaccine-result-was-suppressed.

[127]Scientific Brief: Community Use of Cloth Masks to Control the Spread of SARS-CoV-2, https://www.cdc.gov/coronavirus/2019-ncov/more/masking-science-sars-cov2.html.

As the country still experiences an exponential growth of the outbreak, with more than 2 000 deaths on November 19, 2020, the coronavirus task force holds its first press conference at the White House since July. Donald Trump is not present, but Anthony Fauci praises the "extraordinary scientific advances" on vaccines. He assures the country that the unprecedented speed of development does not compromise their safety. In an interview with NBC, Dr Fauci regrets the lack of transition of presidential power and the lack of coordination between the Trump administration and Joe Biden's team, which "could not only harm the federal coronavirus response at the pandemic's most dire moment, but might also stall the rollout of potential vaccines amid positive medical developments".[128] The previous week, Secretary Alex Azar said the White House's task force would communicate with the President-elect's team only "if and when there's a determination that there will be a transition." This confirms, once more, the extreme politicisation of the pandemic in the United States. For his part, Joe Biden rules out the possibility of national lockdown, thus in line with the position of the task force. In early December, Joe Biden says he is keeping Anthony Fauci on as a chief medical adviser and a member of his Covid-19 Response Team. Just before Christmas, Deborah Birx announces she would soon retire from government service after a 40-year long career in government. Dr Birx has been criticised for having downplayed Donald Trump's suggestion that injecting disinfectant could ward off the virus and for having taken part in a family trip around Thanksgiving despite urging Americans not to travel over the holidays.

In an interview given to the *New York Times* under the new presidency,[129] Anthony Fauci opens up about the years spent advising Donald Trump and regrets the President's reluctance to listen to experts, over the advice of friends and political allies. Although Fauci never considered resigning, he does not have pleasant memories about the meetings at the White House: "The staff would always try to play down real problems and have a little happy talk about how things are OK. And I would always say, "Wait a minute, hold it, folks, this is serious business." So, there was a joke—a friendly joke, you know—that I was the skunk at the picnic."

This politicisation of science and health policies, and even scientific research, backfired against Donald Trump in the final weeks leading up to the election. Pfizer/BioNTech and Moderna were in the race for the messenger

---

[128] Forgey Q (2020, November 16) Fauci warns that White House transition delays could slow vaccine rollout, Politico, https://www.politico.com/news/2020/11/16/fauci-transition-delays-vaccine-rollout-436759.

[129] McNeil Jr. DG (2021, January 24) Fauci on What Working for Trump Was Really Like, The New York Times, https://www.nytimes.com/2021/01/24/health/fauci-trump-covid.html.

RNA vaccine. But Albert Bourla, CEO of Pfizer, refused the funding that the U.S. government was offering in spring 2020 while Stéphane Bancel, CEO of Moderna, accepted nearly USD 2.5 billion as part of Warp Speed. It is true that Moderna is still a small company, a "start-up" with only 800 employees and has not yet placed any product on the market. In January 2020, when the pandemic began, Stéphane Bancel decided to engage his company in the trail of the messenger RNA vaccine. "Here he goes again. He's crazy:" this is how his team reacted to this news in early 2020.[130] "Are you sure we should be doing this?" Dr Stephen Hoge, the company's president, asked him at a February executive committee meeting. The problem is that there is a price to pay for political funding: on August 25, Stéphane Bancel receives a call from the White House: Moncef Slaoui, who coordinates the efforts of the administration on the production of a vaccine, asks him to delay his vaccine production plan. As the United States goes through one of its greatest racial crises caused by police violence against black people, the official believes the company has not recruited enough African and Hispanic volunteers in its clinical trials. He therefore asks Moderna to provide proof of the efficacy of his vaccine for these minorities. Otherwise, insists Moncef Slaoui, the government support will be withdrawn. This explains why, contrary to the President's intentions, the announcement of the first American vaccine was made after November 3, and by the alliance Pfizer/BioNTech…

The history of Covid-19 in the United States constitutes an exemplary textbook case of the almost total marginalisation of scientific and medical expertise in the management of the crisis, as the documentary *Totally Under Control* brilliantly shows. Throughout the pandemic, Donald Trump has taken decisions that are almost the direct opposite of the scientific and medical advice he received. He started downplaying the disease and never seriously addressed the lack of protective equipment and screening tests. This continued with his public support for hydroxychloroquine, his refusal to wear a mask, his repeated calls to reopen the economy, and the restart of his campaign rallies. The President ignored the many recommendations provided by the CDC, especially on reopening the economy. Not surprisingly, the number of cases of infection rebounded in mid-June in half of American states, and then in October across the whole country. Last but not least, Donald Trump did not refrain from showing publicly his disdain for scientific advice, as he was keen to interrupt and contradict his top advisers at the White House briefings. During these daily meetings with the press (until

---

[130]LaFraniere S, Thomas K, Weiland N, Gelles D, Gay Stolberg S, Grady D (2020, November 21) Politics, Science and the Remarkable Race for a Coronavirus Vaccine, The New York Times, <u>Politics, Science and the Remarkable Race for a Coronavirus Vaccine</u>.

the beginning of June), the two doctors were merely filling in the blanks…
This attitude was also reflected on the international scene: the seat that the
United States occupies in the WHO executive board has been vacant for more
than two years and in 2017, Donald Trump cut by more than two-thirds the
CDC staff team based in Beijing, who could have signalled the alert at the
start of the epidemic and improved the cooperation between the U.S. and
China on the coronavirus and the pandemic.

In interviews aired by *CNN* on March 28, 2021, Debora Birx said that
"there were about a hundred thousand deaths that came from the original
surge. All of the rest of them, in my mind, could have been mitigated or
decreased substantially."[131] Robert Redfield, the former director of the CDC,
accused the upper ranks of the Trump administration and in particular the
Secretary of Health and Human Services, Alex Azar, of pressuring him to
change the federal guidelines on coronavirus testing.[132]

The handling of Covid-19 also reflects the fact that Donald Trump has not
considered science a priority.[133] He cut the budgets of many science agencies
and waits almost two years before appointing a science adviser. The role of
scientific expertise has been minimised so as not to thwart the President's
roadmap. But if these shortcomings primarily concern the Trump administration, they have been grafted onto an American society in the grip of
many difficulties. While the United States has one of the highest living standards on Earth, this country faces problems which have been exacerbated
by the crisis and have become very significant, such as underfunding and
the privatisation of the public health sector (which represents barely 15% of
U.S. health spending for a total amount of nearly a trillion dollars), persistent social and racial inequalities, prison overcrowding, excessive promotion
of individualism, omnipotence of social networks and overly "US-centred"
leadership.[134] The President promised in April 2020 that the government
would help those without health insurance, but his words were not followed
by any concrete action. Donald Trump also focused on border and immigration control while their impact on the circulation of the virus is close to
zero. However, what is incomprehensible for Europeans like me is that the

---

[131] Howard J and Kelly C (2021, March 28), Birx recalls 'very difficult' phone call from Trump
following her Covid-19 warnings, CNN, https://edition.cnn.com/2021/03/28/politics/birx-trump-
covid-very-uncomfortable-phone-call/index.html.

[132] The New York Times (2021, March 29) Covid-19: Birx Lashes Trump's Pandemic Response and
Says Deaths Could Have Been 'Decreased Substantially', https://www.nytimes.com/live/2021/03/28/
world/covid-vaccine-coronavirus-cases#cnn-birx-fauci-giroir-interviews-trump.

[133] Tollefson J (2020, October 5) How Trump damaged science – and why it could take decades to
recover, Nature, https://www.nature.com/articles/d41586-020-02800-9.

[134] Yong E (2020, September) How the Pandemic Defeated America, The Atlantic, https://www.the
atlantic.com/magazine/archive/2020/09/coronavirus-american-failure/614191/.

country struggled hard to develop its first diagnostic tests. While Americans have the best health experts, the highest number of Nobel Prize winners and cohorts of innovators envied the world over, this country was clearly unprepared for a pandemic. And the Covid-19 one is clearly not the worst of all! In short, the coronavirus found in the United States a favourable ground for its development.

Some scientists have tried to quantify the impact of the mistakes made by the Trump administration in terms of the number of lives. This is what Michael Riordan, an American scientist who taught the history of physics did. He assumes that the United States has a priori as many resources as Germany to tackle the Covid-19 epidemic, given that these two countries have a comparable level of development and an almost identical gross domestic product (GDP) per capita. On this basis, Riordan estimates that mortality should have been of the same order of magnitude in these two countries, or roughly 114 deaths per million inhabitants (as of September 30, 2020). The balance sheet of Covid-19 in the United States should therefore have been around 38 000 deaths on that date. The difference between this number and the real death toll on that date, nearly 170 000, can be attributed, according to Riordan, to the mistakes and failures of the US administration.[135] Would Donald Trump therefore have close to 200 000 deaths on his conscience? The reasoning is a bit simplistic but it has the merit of quantifying the heavy price paid by the United States to this pandemic. It should also be noted that the same calculation applied to France and the United Kingdom shows that these two countries could have limited, on September 30, 2020, the number of deaths to 8,000 and 7,000 respectively (against 32,000 and 42,000).

On December 9, 2020, the coronavirus is still killing more and more people. The United States breaks the daily record for the number of coronavirus deaths two weeks in a row, as a brutal surge propagates across the country with more than 3 000 new fatalities reported on a single day. The total death toll approaches 300,000, a world record. On December 10, the FDA's independent expert panel voted overwhelmingly (17 to 4) to approve Pfizer/BioNTech's vaccine, given that "the known and potential benefits of this vaccine outweigh the known and potential risks of the vaccine, for the prevention of Covid-19 in individuals 16 years of age and older." The FDA gave the green light on the evening of the next day, December 11, making the United States the sixth country to approve this vaccine, after

---

[135]Riordan M (2020, September 30) The human cost of the Trump pandemic response? More than 100 000 unnecessary deaths, Bulletin of the Atomic Scientists, https://thebulletin.org/2020/09/the-human-cost-of-the-trump-pandemic-response-more-than-100000-unnecessary-deaths/?utm_source=Newsletter&utm_medium=Email&utm_campaign=ThursdayNewsletter10012020&utm_content=DisruptiveTechnology_HumanCost_09302020.

the United Kingdom, Canada, Bahrain, Saudi Arabia and Mexico. A few hours earlier that same day, at 1 p.m. exactly, Donald Trump sent one of those tweets that only he knows how to write: "While my pushing the money drenched but heavily bureaucratic @US_FDA saved five years in the approval of NUMEROUS great new vaccines, it is still a big, old, slow turtle. Get the dam vaccines out NOW, Dr Hahn @SteveFDA. Stop playing games and start saving lives!!!" Donald Trump's presidency ends up in chaos while the epidemic continues to break records.

On January 20, 2021, his first day in office, Joe Biden signs three executive orders to address the challenges of the pandemic. The first order requires masks and physical distancing in all federal buildings, on all federal lands and by federal employees and contractors; the second one aims at coordinating the federal government's response to Covid-19 and involves in particular setting-up a dedicated agency and the third one aims to allow the United States to retain its membership in WHO. Furthermore, the President says he would use federal power to mandate wearing a mask in public. Joe Biden also appoints Jeff Zients to coordinate the President's "Covid-19 Response Team," following the dissolution of the Covid-19 advisory board on the same day, January 20. A member of the Obama administration, Jeff Zients is known in Washington as "Mr Fix it" for his ability to rescue failing government projects. The day after his swearing-in, on January 21, 2021, Joe Biden makes it clear that he wanted to depoliticise the fight against the pandemic: "Our national strategy is comprehensive, it's based on science, not politics. It's based on truth, not denial, and it's detailed," he said. The new President decides to display a portrait of Benjamin Franklin in the Oval Office to show his interest in following science.

On February 22, 2021, the United States passes the grim milestone of 500 000 Covid-related deaths, the highest number in any single country as the Americans' average life expectancy drops by one full year due to Covid-19. However, after the number of new daily deaths peaked around January 12, 2021, the situation is significantly improving and the number of cases dipped by more than 100% in only one month. Beginning of April, 2021, 65 000 new infections are reported on average each day. That is 26% of the peak—with the highest daily average reported on January 8, 2021. With 916 000 jobs created in March 2021 and the unemployment rate falling to 6 percent, the economy is working its way back to pre-pandemic norms as the number of vaccinations continues to rise. There is light on the horizon…

This chapter has shown how arduous the management of Covid-19 has been—and still is. What makes this pandemic so peculiar? Obviously, scientists and politicians have difficulty in cooperating and working together to manage the crisis. Within these partnerships, the alchemy often does not work. Is it because of a lack of trust and dialogue, like in so many ordinary couples? Probably. Is it because SARS-CoV-2 is a new and contagious coronavirus? Sure, although this pandemic is neither the first nor the worst in this century. Or is it because this crisis profited from the many weaknesses of our society? Possibly. The fact is that top politicians and star scientists are all attracted to the glitz of the job. Are they all doing politics? This is what we are going to discuss in the next chapter.

4

Political Distancing

In philosophy, as in politics, the longest distance between two points is a straight line
Will Durant

Criticism is always easy with hindsight. However, in the case of Covid-19, it was from the very start of the pandemic that the critics came down forcefully on most governments. Of course, in Europe in particular, people are quick to criticise their governments. It is an instinctive reflex, in the face of widespread concern to search for the culprits, or at least those responsible who can become scapegoats. Donald Trump did just that, naming SARS-CoV-2 "a Chinese virus," whose threat had been exaggerated by "the media and the Democrats" and cutting ties with WHO, whom he accused of "complacency with China." Let's face it: we have all thrown stones at our governments and politicians haven't shied away from it either. The crisis is not over, yet there have been many reshuffles, resignations and layoffs at the highest level. Donald Trump paid the price by his election defeat. Accused of a lack of preparation, a lack of reaction and a lack of communication, global leaders got a bad press. All of them though? Actually not. In some countries, there are men and women in power who emerge victorious from the crisis. Such as in New Zealand, where Prime Minister Jacinda Ardern has handled the crisis in an exemplary manner, putting in place in less than a month a complete programme for test, trace and isolate. The result: 26 deaths and a mortality of barely 5 deaths per one million inhabitants.

© The Author(s), under exclusive license to Springer Nature Switzerland AG 2021
M. Claessens, *The Science and Politics of Covid-19*,
https://doi.org/10.1007/978-3-030-77864-4_4

However, most of the countries confronted with Covid-19 have shared more or less the same shortcomings in the handling of the crisis: delayed decisions by the authorities, a lack of equipment in hospitals, insufficient testing, etc. A stunning universality in diversity! Of course, no one was really prepared to confront a crisis of this magnitude. Of course, we should not panic and lose all confidence in our leaders forever more, even if we discover during these crises that decision-making is not their strong point. Nevertheless: "gouverner c'est prévoir" (governing means anticipating) Emile de Girardin used to say. From this point of view, governments have failed miserably, with only a few exceptions. The reality is that most leaders were absolutely unprepared. They faced the crisis with artisanal means, with a high degree of improvisation, often ignoring their own administration despite it being well established that a pandemic constitutes the highest health risk in our societies and, as such, requires a prevention and protection strategy worthy of the name. Strengthening health systems is a national defence priority. Despite the losses caused by recent AIDS, SARS, Ebola and Zika outbreaks, our leaders were unable to project that a virus that appeared in an unknown city in China could cause so much damage in their own country and on all five continents. "The response of governments to Covid-19 represents the greatest political failure of Western democracies since the Second World War," writes Richard Horton, editor-in-chief of The Lancet and author of one of the first books on this pandemic.[1] He adds: "To blame China and WHO for this global pandemic is to rewrite the history of Covid-19 and to marginalise the failings of Western nations." In *Le Monde*, journalists Hervé Morin and Paul Benkimoun also conclude: "This pandemic is a disaster that we created ourselves.[2]"

There is nothing new under the sun: what happened in the west had already happened elsewhere and solutions had already been found. Thus, the comparison of the cities which, during the influenza pandemic of 1918–1919, adopted strict public health measures (such as closing schools, banning public meetings, imposing quarantines, etc.) with those which opened more quickly and adopted fewer restrictions shows that the former endured fewer losses and experienced a faster economic recovery.[3] Arguably, in the heart of

[1] Horton, R., The Covid catastrophe, Cambridge, Polity Press, 2020.

[2] Morin H and Benkimoun P (2020, June 20) Le Covid-19 montre une faillite catastrophique des gouvernements occidentaux, Le Monde, https://www.lemonde.fr/sciences/article/2020/06/20/richard-horton-le-Covid-19-montre-une-faillite-catastrophique-des-gouvernements-occidentaux_6043590_1650684.html.

[3] Correia S, Luck S, Verner E (2020, June 5) Pandemics Depress the Economy, Public Health Interventions Do Not: Evidence from the 1918 Flu, Correia, Sergio and Luck, Stephan and Verner, Emil, Pandemics Depress the Economy, Public Health Interventions Do Not: Evidence from the 1918 Flu (June 5, 2020), available at SSRN: http://dx.doi.org/10.2139/ssrn.3561560.

the crisis, our governance methods have to cope with a stark reality. There is a huge gap between the policy ideas and the operational management of an epidemic. Since December 2019, epidemiologists have been pushed to change their working methods from traditional research to instant advice.

In this chapter, we look at the context in which the pandemic developed, that is modern society at large. We look in particular at how our political leaders interact and communicate with the scientific community in order to design and implement research priorities and public health policies. This will allow us to establish the necessary conditions for genuine cooperation and mutual respect, which are all the more indispensable to set up a long-term international interaction between politicians and scientists and tackle a planetary crisis like this one.

Thousands of Excess Deaths

During the first half of 2020, in the 196 countries and territories hit by Covid-19,[4] repetitive bad news circles were the rhythm of the pandemic. In theory, humanity has never been so well prepared to face a new pandemic, with its scientific knowledge, its sophisticated technology, its pharmaceutical industry and its epidemiological models. But the management of the health crisis by the countries on the frontline is confused—and confusing. In short, the way governments have communicated in these countries during the pandemic has been a catastrophe. "I will not let anyone say that there was a delay in taking the decision on lockdown," French Prime Minister Edouard Philippe said during a press conference on March 28, 2020. Nevertheless, the facts give a very different view of things. In France, the United Kingdom and the United States, the surge of Covid-19 has been accompanied by a denial of any delay and of any political responsibility, as we have seen.

But how much delay are we talking about? Table 4.1 summarises the situation in six of the countries that were on the frontline during the first wave of spring 2020: China, France, Italy, Spain, the United Kingdom and the United States. I have added Greece, which still has one of the lowest Covid-19 death rates in Europe, the government of Kyriákos Mitsotákis having quickly realised that time is our number one enemy. The same can be said for Vietnam, which reacted immediately on January 10, i.e., the day after the announcement of the first Covid-19 death in China, by imposing control

---

[4]The 194 world countries, plus 2 international territories (the Diamond Princess and MS Zaandam cruise ships).

Table 4.1 Key dates for the first wave of the Covid-19 epidemic in frontline countries

	China	France	Italy	Spain	United Kingdom	United States	Greece
First official case	December 30, 2020	January 24, 2020	January 31, 2020	January 31, 2020	January 31, 2020	January 21, 2020	February 27, 2020
Lockdown	January 23 (Wuhan)	March 16	March 10	March 15	March 23	March 30 (30/50 States)	March 23
Delay between first case and lockdown (weeks)	3.5	7.5	5.5	6.5	7.5	10.0	3.5
Lifting of lockdown	April 8 (Wuhan)	May 11	May 4	April 14	June 1	?	May 4
Covid-19 mortality (number of deaths per 1 000 000 inhabitants)	78 (Hubei)	1600	2100	1700	1900	1800	1200

checks at the borders, closing schools and applying strict isolation. Result: 66 deaths for 13 258 confirmed cases.

I do not mean here that there is a direct relationship between the Covid-19 mortality and the time to lockdown. National and local situations are obviously complex. However, France, the United Kingdom and the United States took longer than China to impose their first lockdown (please remember that the U.S. never imposed a federal shutdown). Greece locked down in 3.5 weeks, like China, and Italy in 5.5 weeks. But given the very rapid increase in the number of infections in the latter country, it is reasonable to believe that the epidemic was spreading long before the official appearance of the first cases. Let us remember the statements of Doctor Paglia.

The case of Vietnam is interesting and reminds us that going into lockdown was not inevitable. However, this very populous (98 million inhabitants) and relatively poor country was not well placed to face a pandemic, on paper. But the government very quickly implemented the "test, trace, isolate" recipe followed by their Asian neighbours and, later, by the rest of the world, as well as a public information policy to which the population adhered. There are many ways to achieve physical distancing, but time has been the primary enemy everywhere: the countries that have performed the best are those that reacted quickly.

The table also recalls that after China, the epidemic broke out almost simultaneously in other countries at the end of January, despite the geographical distance, a result which shows that international travel made a significant contribution to the spread of the coronavirus.

Why have most countries been so slow to take the measure of the epidemic—and therefore so slow to implement social distancing measures, the only tools capable, during those early weeks, of containing or even halting the epidemic, despite China's full lockdown and WHO's highest level of alert? The leaders had all the information and simulations necessary, showing that, at the start of the epidemic, a delay of one week could double the number of infections and deaths. We have mentioned in the previous chapters two scientific papers which show that the epidemic was spreading silently in Europe at the end of 2019. As explained in the first chapter, the characteristics of the coronavirus and the epidemiological data (high contagiousness, long period of incubation and asymptomatic cases) predicted an early exponential expansion that would sooner or later explode without non-pharmaceutical interventions. As early as January 2020, Italian researchers in Marcello Tirani's team concluded: only "aggressive" strategies and decisions taken without delay could have stopped the spread of the virus. This is not what happened. Nonetheless, Marcello Tirani confirmed to me that the Italian government took some of their conclusions into account. In France, a study by the Ecole des Hautes Etudes en Santé Publique concluded that 73 900 people would have died in hospital between March 19 and April 19 in the absence of any control measures, i.e., six times more than the 12 200 deaths observed over this period. Even if some of the study's assumptions are questionable, the researchers conclude that the lockdown saved at least 62 000 lives in France that is an 84% reduction of the number of deaths during that period.[5] "Pure fantasies," Jean-François Toussaint concluded, a professor at the University of Paris.

The addition is obviously even heavier for countries which have moved slowly and belatedly towards lockdown, such as the Netherlands, Mexico, Brazil, the United Kingdom and, of course, the United States, which never imposed a nationwide shutdown. The results are also worsening in Sweden, which adopted an approach somewhat different from the other EU countries, at least until November 2020. While its Scandinavian neighbours (Denmark, Norway and Finland) decided early on to go into lockdown, this was not the case in Sweden. This country, which has one of the lowest population

---

[5]J. Roux, C. Massonnaud, P. Crépey, Covid-19: One-month impact of the French lockdown on the epidemic burden, EHESP, 23 avril 2020, https://www.ehesp.fr/wp-content/uploads/2020/04/Impact-Confinement-EHESP-20200322v1-1.pdf.

densities in Europe (22 inhabitants per square kilometre, against over 400 in the Netherlands for example), adopted very light restriction measures until the end of 2020, banning only gatherings of more than 50 people, visits to retirement homes and bar service in cafes and restaurants. Nurseries, primary and secondary schools, swimming pools and sports halls remained open. Telecommuting was only recommended. The mask is not compulsory anywhere. Authorities called for individual responsibility. Swedes have been urged to keep their social interactions to a minimum, but the government nonetheless encourages them to exercise and leave their homes. This is what the Swedish call "freedom under responsibility." History will tell if, in the end, this strategy has paid off, but during the first months of the epidemic, it was well received by the population even if Swedish Prime Minister Stefan Löfven declared on April 3, 2020 that his country could be facing "thousands of coronavirus deaths" and that the crisis was likely to drag on for months rather than weeks. At the lowest polls in February 2020, the government was able to regain the confidence of the Swedes, 63% of whom had a positive assessment of the handling of the crisis on April 21. This may be due, at least in part, to the fact that the executive admittedly based its strategy on the recommendations of scientists such as epidemiologist Anders Tegnell, who has become a well-known and popular figure in the country since the start of the pandemic. However, the statistics do not play in their favour. And Sweden has also been hit by the economic crisis much like their neighbouring (and confined) regions. Then the experts started to criticise their government's strategy of dealing with the pandemic: in a column published in the newspaper *Dagens Nyheter* on April 14, twenty-two doctors and scientists asked their political leaders to drastically change their method and strengthen physical distancing. In early November 2020, the country saw a significant surge in the number of cases and the government had then no alternative but to change tack: Stefan Löfven decides to ban the sale of alcohol from 10 pm on November 10, forcing bars and restaurants to close at 10:30 pm. On November 16, the Prime Minister announces a ban on public gatherings of more than eight people and implores the Swedes to stay at home. In early May 2020, Sweden had a mortality of 400 deaths per million inhabitants, several times more than its neighbours like Norway and Finland. And the numbers only got worse. The country has so far resisted going into lockdown, unlike the rest of Europe, even during the peak of its second wave over Christmas, but Sweden has now one of the highest Covid-19 mortalities in Europe, about ten times higher than Norway and Finland.

The effects of NPIs are nevertheless confirmed by a recent study based on international comparisons, albeit with more nuanced conclusions.

Researchers from Stanford University have shown that NPIs led to significant reductions in the growth of new cases in 9 out of 10 countries studied (England, France, Germany, Iran, Italy, Netherlands, Spain, South Korea, Sweden and the United States) during the spring of 2020.[6] Only Spain does not demonstrate a significant effect. However, the researchers did not find significant benefits in the hard-hit countries which implemented more restrictive NPIs such as mandatory stay-at-home orders and business closures i.e., all the ten countries minus Sweden and South Korea.

Of course, we all know that we should be careful when comparing one country to another. The impact of the coronavirus in a group, in a city or in a country depends on a multitude of factors. The number of deaths is in principle an objective measure but the official figures generally provide an underestimate of the reality since, in some countries, deaths outside of hospital are neither tested nor counted (in particular in the United States, where the rules vary from one jurisdiction to another; or in Russia, where some Covid-19 deaths from have been counted as pneumonia and therefore do not appear in the official statistics), not to mention the fact that some governments seek to minimise the impact of the pandemic on their territory. However, the balance sheets are refined as readjustments are made.

I always find it disappointing when the media write headlines about the absolute number of cumulative deaths in a given country. This denotes an obvious lack of mathematical and statistical understanding by the authors. What do 60 000 deaths mean for a country like Brazil, which has 209 million inhabitants? What do 10 000 deaths represent for a small country like Belgium (the size of Wuhan)? Intuitively, we can see that the Belgian numbers are worrying. In reality, the only relevant indicator here is the number of deaths reported compared to the total population, that is the Covid-19 mortality, as it takes into account the demographic size of each country. The same is true for all variables that depend on the size of the system, such as the wealth of a particular country. The richness of a nation, in absolute terms, has only an academic interest. On the other hand, it is the GDP (gross domestic product) divided by the number of inhabitants that makes it possible to compare national performance.

Figure 4.1 shows significant differences in this indicator between countries. The comparison with China is particularly revealing. Thus, France has, in relative terms, 450 times more deaths than the Middle Kingdom, the United Kingdom and the United States respectively 550 and 500 times more.

---

[6]Bendavid E et al. (2021, January 5) Assessing mandatory stay-at-home and business closure effects Assessing mandatory stay- at- home and business closure effects on the spread of COVID- 19, Eur. J. Clin. Invest. 2021;51:e13484, https://doi.org/10.1111/eci.13484.

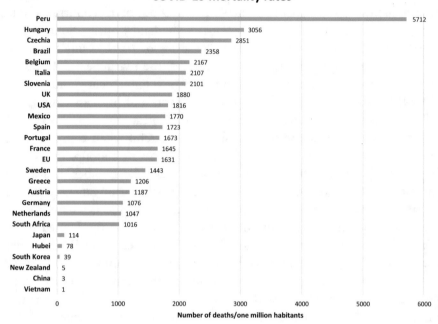

Fig. 4.1 Number of Covid-19 deaths per one million inhabitants. Data as of June 21, 2021, Source: Johns Hopkins University

The interpretation of Fig. 4.1 is not straightforward. Nevertheless, major trends are appearing. First, we see a "Western" bloc and an "Asian" bloc, which may signal the role of geographic and/or cultural factors. Then countries that have been slow to impose tough measures, such as Brazil, the United States, the United Kingdom, Sweden and the Netherlands tend to be on top of the chart. Conversely, countries that quickly imposed containment or a strategy for the identification and isolation of infected people, such as Austria, Germany, Greece, South Korea, Japan, China, New Zealand and Vietnam are lower down on the list. Austria's relatively good rate shows that even in Europe, governance based on common sense and fear, motivated by Chancellor Sebastian Kurz's "Danger is among us" phrase, can pay off. But above all, these eight countries have shown that a rapid reaction, from the first symptoms, pays off. These data also confirm that there are several methods to effectively achieve social distancing. Lockdown is not inevitable.

The combination of Fig. 4.1 and the testimonies of the previous chapters confirm the failure of outbreak management in many countries. In France, the United Kingdom, the United States and many other countries, governments, by almost systematically ignoring warning signals or waiting too long to react, have paid a high price. The examples compiled in this book show that these governments consciously, continuously and deliberately delayed taking the actions that they knew could have saved many lives. If there is one lesson to be drawn from the Covid-19 pandemic it is this: speed is everything. By underestimating the risks of the epidemic, the leaders of these countries have placed a huge burden on their respective populations. There is no doubt our leaders expected to find an easy way out of the crisis; no doubt they were afraid of giving in to panic; no doubt they tried to save the economy while protecting the people. But officials were weak and policymakers have failed. The then future President of the United States, Joe Biden, did not say anything else on March 12, 2020: "A tragedy is added to the crisis, it is all the suffering that could have been avoided thanks to a quick decision and to decisive action. It is a challenge that requires leadership. It requires transparency and urgency." He also added that "the nation must be guided by science." Furthermore, several of those countries that have accused China of withholding information—a classic political tactic aimed to shift the focus—have shown an incredible lack of transparency at home. It is a global political scandal.

The toll is even heavier in retirement and nursing homes in most of these countries. In Europe, the loss of life has been overwhelming: almost half of the people who died from Covid-19 are residents in long-term care facilities. "An unimaginable human tragedy," Hans Kluge declared on April 23, 2020. Many families pointed out the lack of tests, care and support endured by their loved ones. Consequences will follow: legal actions are increasingly being filed in European countries. In France for example, dozens of complaints have been made, in particular against the State, against X or against managers of retirement homes for reasons including poor management of the health crisis, non-assistance to a person in danger, endangering the health of other people, lack of testing, and even crime (involuntary homicide), etc. It is an incredible public health scandal. It had already been established in January 2020 in Wuhan that the coronavirus mainly killed the elderly. A French website and magazine aimed at seniors wrote in black and white at the end of January 2020, so at the very early stages of the epidemic that: "The average age of people who have died from this coronavirus is 70 years old. It's no surprise therefore that the elderly are the main victims of this epidemic.[7]" There has been a culpable delay in taking into account the situation of nursing homes for which the responsibility falls both on the government and on those in

---

[7]Senioractu (2020, January 29) Coronavirus de Wuhan: la plupart des morts sont des personnes âgées…, Senioractu, https://www.senioractu.com/Coronavirus-de-Wuhan-la-plupart-des-morts-sont-des-personnes-agees_a22435.html.

charge of these specialised centres. Anthropologist Jean-Dominique Michel confirms it: "The conclusion of this sad story is that the losses (human) and damage (material) that we have suffered are mainly due to poor responses from the authorities which have aggravated a health problem which was very much controllable".[8]

Leaders across the world—but in particular in France, the UK and the US—were nevertheless advised by scientists and doctors. Yet in the end, they failed to control the risks of the pandemic and the relentlessness of the geometric sequence. No doubt they listened to the experts, but the resulting sequence of events seems to indicate that this advice was not really heard. Are these scientific committees just paying lip service then? Were their opinions unconvincing? Have economic and social arguments taken precedence over the health emergency for too long? Still, experts failed to influence governments early and decisively. We will come back to these questions in the following pages. If our leaders sometimes tend to place themselves above the law, they cannot place themselves above science.

This is also the conclusion of Richard Horton: "Why did President Macron, President Conte, Prime Minister Johnson, President Trump do nothing? Did they not understand what was going on in China? Did they not believe the Chinese? Didn't they ask their diplomatic representations in Beijing to investigate? I do not understand. The evidence was very clear by the end of January. So, I think politicians are going to have to explain themselves.[9]"

At the global level, these delays in political decisions, obvious in many countries, seem incomprehensible. And the delays were not just about lockdown. There have been reversals and inconsistencies about mask-wearing, testing, drugs, restrictions, etc.in virtually every country except Asia. When the epidemic resumed in France on July 14, the government and health authorities were content, for the most part, to monitor and comment on the evolution of the statistics. Early March 2020, two weeks before lockdown, Boris Johnson was still advocating a *laissez-faire* Covid-19 strategy and recommended only regular hand washing. A few weeks before the end of his mandate, Donald Trump declared that "the disease is already disappearing" and is no longer threatening America's health and the nation's economy.

---

[8]Michel JD (2020) Covid: anatomie d'une crise sanitaire, humenSciences, Paris.

[9]Morin H and Benkimoun P (2020, June 20) « Le Covid-19 montre une faillite catastrophique des gouvernements occidentaux», Le Monde, https://www.lemonde.fr/sciences/article/2020/06/20/richard-horton-le-Covid-19-montre-une-faillite-catastrophique-des-gouvernements-occidentaux_6043590_1 650684.html.

In the light of the testimonies presented above, these delays appear to result mainly from a collective downplaying the severity of the crisis. In China, the decision to shut down Wuhan, Hubei, and then all of China meant huge economic loss. But it had become clear to Chinese leaders that avoiding lockdown and managing the situation by other means would have created an even greater loss. Since the summer of 2020, life in China has almost returned to normal. But vigilance is key: when three new Covid-19 cases are discovered on October 11, 2020 in the city of Qingdao (nearly nine million inhabitants), the authorities immediately decide to test the entire population in three days. It is indeed at the very beginning of an exponential process that one can control it relatively easily; later, only radical means can achieve that.

Fearing legal action, many politicians today seek to discharge their responsibilities onto officials, experts and other scapegoats. Fully aware of the dangers of the epidemic and of the effective measures taken by China in Hubei, leaders in Europe and America today have thousands of deaths on their conscience. It is little consolation that in the response to SARS-CoV-2, democracies show mixed results but populist regimes failed "even more than the others," according to the American political scientist Yascha Mounk.[10]

Beyond this grim record, let us not lose sight of the fact that several countries successfully managed the crisis, without going into full lockdown, such as South Korea, Vietnam, Taiwan, Singapore, Germany, Austria, Switzerland and many more. There are indeed three strategies for reducing the reproduction rate of a viral epidemic (in the absence of vaccines and antiviral remedies):

- Minimising social interactions (for example through lockdown and banning meetings) as most European countries have done, with mixed success;
- Ensuring that our contacts do not transmit the disease (barrier gestures, hygiene measures, wearing face masks, etc.)—the typical examples here are Germany and Sweden (although the latter show much less convincing results); and
- Testing and isolating, to prevent infected individuals from coming into contact with other people during contagiousness (isolation, self-quarantine, etc.) as in China, South Korea, Taiwan, etc.

---

[10]Georges B (2020, April 22) « Nos démocraties deviennent incapables d'affronter des crises», Les Echos, https://www.lesechos.fr/monde/enjeux-internationaux/yascha-mounk-nos-democraties-deviennent-incapables-daffronter-des-crises-1196997.

South Korea was hit hard by the virus very shortly after China. The government quickly implemented an intensive control and prevention programme. Using personal data from cell phones, the authorities were able to trace infected people and their contacts, and isolate them if necessary. In addition, anyone could request a Covid-19 test if they wished. As a result, the epidemic was brought under control within a few weeks and health professionals managed to prevent epidemic rebound. The country now has less than two thousand deaths and one of the lowest mortality rates in the world. The strategy followed by Singapore was roughly the same but did not initially achieve the expected success: after a period without a strict quarantine, increasingly restrictive measures were imposed, until total lockdown came into force on April 7, 2020. Despite a very efficient mobile phone tracking system, the country had to face a second wave of infection and on April 22 decided to return into lockdown until June, 2020. The death toll is as low as 30 deaths. Switzerland has also ruled out any strict lockdown even if, subsequently, the country has toughened its measures against the coronavirus. Swiss Health Minister Alain Berset said that, "[unlike other countries], we do not play politics."

Herman Goossens, a microbiologist at the University of Antwerp and coordinator of a European project on emerging epidemics, now believes that the first wave could have been better managed in Belgium. He regrets that, unlike Taiwan for example, his country was not carrying out mass testing and mandatory isolation as early as January 2020. "A colossal error", the scientist believes, "that most other European countries have made, except Germany. I am frustrated about this. Not because of Belgian politicians or because of my fellow experts, but because of the system. Investigation and hearings will have to be organised to understand how is it that we did not receive any information about what was going on in Taiwan?[11]" The statements of epidemiologist Arnaud Fontanet, member of the French scientific council, to the Parliamentary Commission of Inquiry of the National Assembly on the management of the Covid-19 crisis, on June 18, 2020, go in the same direction: "The only countries which have succeeded in controlling the epidemic without containment, have done so thanks to a whole series of measures including a system for efficient tracing and a culture of wearing masks and hand hygiene.[12]"

---

[11]Rigot M (2020, May 16) Coronavirus en Belgique: « Nous avons commis une erreur colossale», La Libre Belgique, https://www.lalibre.be/planete/sante/coronavirus-en-belgique-nous-avons-commis-une-erreur-colossale-5ebf8802d8ad581c54e696b8.

[12]Le Monde avec AFP (2020, June 18) « On n'avait pas le choix»: des membres du conseil scientifique défendent le confinement, Le Monde, https://www.lemonde.fr/politique/article/2020/06/18/on-n-avait-pas-le-choix-des-membres-du-conseil-scientifique-defendent-le-confinement_6043328_823448.html.

However, it is too simplistic to say that the countries that have controlled or eliminated the virus did things extremely differently. The country that had the strictest lockdown in the world is Peru, and they are devastated. The fact is that neither France, nor the United Kingdom, nor the United States, nor many other countries took the measure of what was happening. They ignored those who really knew what was going on. It is the tragedy of all societies, including democracies: those who hold power tend not to give up any of that power. Too often the primary objective of leaders is to establish, reinforce and publicly exhibit the power they hold in their hands. And that goes for all elites, including the scientific and medical communities. But in all of these countries too, those who knew were not clear enough. In reality, the failure is a shared one because we all have, more or less, a responsibility, both individual and collective, in the inability of our countries to contain the pandemic. One just needs to realise that, in most countries, the epidemic rebound at the same pace as vacations, holidays and celebrations...

The only real question left hanging is this: could a "Chinese-style" Covid-19 strategy be implemented in democratic countries? The choice between three months of very harsh lockdown, genuinely "inhuman" from several points of view, and more than one year of successive waves and lock-downs with serious psychological consequences and heavy economic impact, which is the "least worse" solution? Inspired by the exemplary successes in the management of the pandemic achieved in Asia and in the Pacific, a growing number of experts and organisations, particularly in the United States, Germany, the United Kingdom and Japan, are advocating a "zero COVID" strategy, in other words the pure and simple elimination of the virus. This idea is championed by Yaneer Bar-Yam, who chairs the New England Complex Systems Institute (NECSI) in Cambridge, USA, and heads the international "EndCoronavirus" coalition.[13] He proposes to achieve this in hot spots through four to six weeks of strict lockdown, travel restrictions, mass testing, contact tracing and rigorous isolation. The health benefit is not the only argument put forward as the economic cost of lockdowns is huge and increases with their duration.

The Emergence of "Science Politics"

In most countries, governments called on experts from the scientific and medical professions to provide advice on the management of the health crisis

---

[13] https://www.endcoronavirus.org/.

and guidance for public action. However, the course of events and the available evidence raise doubts about the quality and/or the usefulness of the exchanges between the scientific and political spheres. Yet the scientific expertise was of a high level: the opinions of the Covid-19 scientific council in France and the minutes of the meetings and the documents of the Sage committee in Great Britain testify to the quality of the information provided to governments and the public. Donald Trump was also very well advised. Yet in France and the UK, against the advice of their experts, governments refused to impose a second lockdown in September. However, they were forced to do so a few weeks later. In the United States, Donald Trump has never supported mask-wearing and has always refused a call on national lockdown. Yet he could rely on the CDC and the best health experts in the world. Only China seems to have fully followed the recommendations of its specialists. What got stuck?

This is nothing new: health crises, global warming and technological accidents pose and re-pose the same question: why are the experts whose task it is to advise the government not getting through? Of course, the final decision rests with the highest political authorities. But why, on the subject of global warming for example, have leaders not yet changed course given the glaring scientific evidence? There is nothing new under the sun, I dare to say! Should we therefore resign ourselves into inaction and accept once and for all that, in the political world, 'doing nothing' is in fact a well-thought-out strategy? "Above all, do not decide anything," concludes Pierre Conesa in his survival guide dedicated to people working in a political environment. Have our leaders become "non-decision-makers"?

In defence of politicians, we should acknowledge the fact that the experts' way of communicating is sometimes very technical, full of jargon and punctilious. Scientists spend most of their time communicating primarily with their peers, in many different forms. They are used to preaching to the choir.

Let me open a small parenthesis here. Given the notorious difficulty in getting scientific messages across in political and media arenas, I like to ask my students or scientists at conferences this question in a straightforward way: is science basically *incommunicable*? I do not mean here that we cannot *communicate* about science—just that we cannot *discuss* science. Science is science and the facts are the facts. There is nothing to argue about an equation such as $E = mc^2$. In a small book recommended by the French CNRS,[14] I show that science, in the strictest sense, is not *communicable* but that scientific knowledge nonetheless disseminates to the public. In reality, the average

---

[14]Claessens M (2009) Science et communication: pour le meilleur ou pour le pire ?, Quae, Paris.

scientific culture is much better than we think. I do not know of any scientific subject that is not accessible to the general public. Of course, the public communication of science is important but some scientists are still keen to shirk any responsibility by saying that this is not among their priorities. End of parenthesis.

There is hardly any doubt that, during this pandemic, politicians have been listening to "their" scientists. But have scientists heard "their" politicians? We have seen that many experts played down the crisis and did not raise the alarm in time.

Obviously, setting up a scientific advisory committee is not a trivial decision: its (relative) success depends on its members, knowing that they might also evolve or reveal a different profile on the political scene and under the spotlights. Experts can mutate just as much as viruses can! The public is not fooled and is quite concerned about the metamorphosis of scientists into experts mandated by public authorities to assess risks and make recommendations. Opinion polls carried out in France, Europe and the United States have clearly shown that, when it comes to estimating risks related to health and the environment, the level of public confidence depends on the distance perceived between scientists and politicians. In France for example, 87% of the people interviewed during the major national poll carried out in 2011 said they "trusted science", but only 48% trust the government agencies responsible for monitoring risks related to health and the environment.[15] Recent studies have confirmed that society's goodwill can decline over specific scientific issues that are highly politicised.[16]

Governments choose their experts on scientific grounds but also according to other criteria such as political affinity, media visibility or just personal connections. The same goes for the expert, who will accept the government's proposal or not, for similar reasons. And the chemistry doesn't always work. We can even say that, once in the political arena, science becomes less *scientific*. Didier Raoult experienced this, after joining the Covid-19 council at the request of President Macron. On March 16, just five days after the expert committee was established, the magazine *Marianne* published an interview with the renowned scientist, in which he declared: "I tell [the council] what I think, but I see it is not translated into action. This is called a scientific

---

[15]Marcé C and Boy D (2012) Les représentations sociales de la science et de la technique, SciencePo, CEVIPOF/CNRS, Paris.

[16]Bauer MW, Pansegrau P, Shukla R (2018) The cultural authority of science: comparing across Europe, Asia, Africa and the Americas, Abingdon, Routledge, https://doi.org/10.4324/978131516 3284.

council, but it is a political one. I am here like an alien.[17]" Professor Raoult then decided, after two meetings, to no longer participate in the Council. In an op-ed,[18] he explains his decision, which is clearly non-scientific: "It is time for physicians to take their place again, with philosophers and with people having a humanist and religious inspiration, in a moral—or *ethical* if we like— reflection, and that we get rid of mathematicians, who are like meteorologists in this field. [...] This explains why I did not want to continue to participate in the scientific council, in which there were two modelers of the future (who, for me, represent the equivalent of astrology), maniacal about methodology. Doctors faced with the problem of providing care represented a minority who are not necessarily used to expressing themselves and who were overwhelmed in this pseudoscientific context. Finally, there is a conflict of interest between becoming the spokesperson for the government strategy and chairing the ethics committee." But the real reason may lie elsewhere: in an opinion dated April 2, 2020, the members of the scientific council criticised their colleague Didier Raoult (without naming him), recalling that "in the epidemic context, researchers and all research stakeholders are required to comply with French and international regulations governing public and private research, in particular in the field of clinical trials.[19]" As we have seen, clinical trials on hydroxychloroquine raised political concerns and ethical issues in France (Jean-François Delfraissy is also the President of France's National Consultative Ethics Committee).

The media love scientific committees. Their members are respected figures, they generally give well-informed advice and some are in tune with the media, or at least understand the rhythm of the media cycle. But some mainstream journalists do not seem to understand the complexity of the job as these experts, aware of their strong public exposure, have to take part in political decisions while at the same time keeping a good distance from the centre of power. In doing so, they also blur the image of science and scientists. Crises are in this respect extra-ordinary because they highlight, under the spotlight of the media, the frontiers of knowledge, which are quickly embodied in the form of dissonance or even discord between the experts who will come

---

[17]Perrier B (2020, March 16) Didier Raoult sur le coronavirus: "Il ne faut pas jouer avec la peur", Marianne, https://www.marianne.net/societe/didier-raoult-sur-le-coronavirus-il-ne-faut-pas-jouer-avec-la-peur.

[18]Raoult D (2020, March 25) « Le médecin peut et doit réfléchir comme un médecin, et non pas comme un méthodologiste», Le Monde, https://www.lemonde.fr/idees/article/2020/03/25/didier-raoult-le-medecin-peut-et-doit-reflechir-comme-un-medecin-et-non-pas-comme-un-methodologiste_6034436_3232.html.

[19]Avis du Conseil scientifique Covid-19, 2 avril 2020, Etat des lieux du confinement et critères de sortie https://solidarites-sante.gouv.fr/IMG/pdf/avis_conseil_scientifique_2_avril_2020.pdf.

to speak on the subject. Scientists, or more exactly experts, then appear in the media as rivals, clashing over their respective approaches and interpretations. At the boundaries of science, discussions may become unscientific. On August 1, the French infectious disease specialist Eric Caumes raises the possibility of allowing young people to infect each other in order to participate actively in the development of herd immunity. The next day, he is contradicted by his colleague epidemiologist Catherine Hill, who recalls on the *TF1* channel that the virus is actually circulating ad hoc without specifically selecting its target. And on August 3, the head of emergency services at the Bichat-Claude Bernard hospital, Enrique Casalino, finds it puzzling that an exception to collective responsibility should be considered. Eric Caumes advanced a hypothesis; Catherine Hill recalled the actual state of play, and Enrique Casalino made a moral or even philosophical point. It is unfortunate that this exchange, which, in the eyes of the public, looks like a quarrel between scientists over "scientific" interpretations, makes headlines in the media. It only generates misunderstandings as conjectures formulated by scientific or medical authorities sometimes turn into pseudoscientific certainties, at least in the eyes of the public. People do not realise that in this case, it is research, not science, that is being publicly communicated.

Hence, the public may be bewildered when science does not speak with one voice. If science is synonymous with fact, how can there be such disagreements? Since the beginning of the pandemic, everyone now has access to daily insights into the realities of scientific research and there is evidence that they don't like what they are seeing. This is why sociologist Michel Dubois, a research director at CNRS, wonders about the impact of Covid-19 on the confidence that the French have in science and technology.[20] He refers to the work of Roger Pielke, who distinguishes between four types of expert which correlate with the way we think about the relationship between science and politics: the "pure scientist" communicates the most 'up-to-date' knowledge devoid of political context; the "science arbiter" responds in a neutral and factual manner to questions raised by the authorities without revealing his own preferences; the "issue advocate" is a kind of lobbyist who defends a precise line and tries to influence the authorities' decisions accordingly; and finally the "honest broker" puts on the table all the possible alternatives so that politicians can make a selection on the basis of their objectives and values. Here lies the crux of the matter, according to Michel Dubois: "The real and lasting impact of the current coronavirus crisis on the image of scientific expertise will also be dependent on the model which will be adopted by

---

[20]Dubois M (2020, April 1) Le coronavirus peut-il altérer la confiance en la science ?, CNRS Le journal, https://lejournal.cnrs.fr/billets/le-coronavirus-peut-il-alterer-la-confiance-en-la-science.

scientific councils and advisors to guide their own action." On this point, let us note that a survey conducted in Germany on April 15 and 16, 2020 (therefore during the Covid-19 pandemic) shows a significant leap in confidence in science and research (73% of people questioned are confident or very confident, compared to 46% in 2019).[21] And surveys carried out in the United Kingdom and the United States during the pandemic but before vaccines were approved confirm the fact that people tend to feel more positive about science and scientists.[22]

At the frontiers of science and politics, these advisory boards are therefore an essential link and a place of communication. But they cannot personify pure and absolute science, whose myth persists in society. The appointment of their members is generally the subject of secret negotiations and often offers up some surprises. Obtaining the chairmanship of a committee is, for a scientist, the assurance of obtaining an exceptional visibility and credibility. We can therefore imagine that these decisions are accompanied by manoeuvres behind the scenes. These committees are therefore a conjunction of knowledge, personalities and interests. Nonetheless, they constitute one of the resources at the disposal of political powers to take into account the latest scientific knowledge and to orient their political decision "scientifically."

The fact remains that, in the case of Covid-19, scientists and doctors cannot boast of any success. Some of those who advised governments failed to have a decisive influence on the timing and the direction of decisions. Could this be down to a lack of firmness or confidence on their part? Is it because scientists are known to be reserved and place great importance on fact-checking? Or is it because, having seen that politics is what it is, they seemingly resigned themselves to having little impact? In any case, in the several countries we have examined, some advisors, albeit among the most competent and the best informed, minimised the gravity of the crisis. Were they afraid of making a mistake? Or afraid to tell the truth? More worryingly, as we have seen, experts themselves acknowledged that they did not rise to the challenge. Part of the explanation is also to be found in cognitive biases, for example the tendency we have to wait for others to act rather than acting ourselves and to seek evidence confirming our own "truth," and

---

[21] https://www.wissenschaft-im-dialog.de/en/our-projects/science-barometer/science-barometer-special-edition-on-corona/.

[22] Jensen EA, Kennedy EB and Greenwood E (2021, March 2) Pandemic: public feeling more positive about science, Nature, Nature 591, 34 (2021), https://doi.org/10.1038/d41586-021-00542-w.

in the fact that most of us prefer a political and social status quo rather than participating in any radical response.[23]

In France, Philippe Juvin, head of emergencies at the Georges-Pompidou hospital in Paris and a member of the Republican Party, casts an enlightening eye on the nature of the dysfunctions, which according to him are deep-rooted and collective: "If the government has failed, many of us have also not realised the magnitude of the health crisis that was going to occur. Although "governing is anticipating," this is a real challenge in this case. Then, there is a real fundamental question, which arose well before this crisis. The experts, the "advisers to the princes" and the plethora of senior officials are not all fully qualified for the work required. Sometimes they do not bring in anything relevant for politicians. They tend to be conservative and they never question themselves. However, we are in an emergency. The chiefs and the deputies, who are used to issue orders and counter-orders, will be held accountable for their actions after the crisis. The personalities who advise the Heads of State must be replaced. This crisis does not happen by chance. The people we pay to anticipate the future have failed. We are relying on people who did not rise to the task that was before them. The alerts were not escalated. The analysis will have to be done after the crisis, but it can already be said that certain administrative rules must be rewritten. The administration must be reorganised. The departments are jostling for power, and are complicating the life of each other.[24]"

As we have seen, on March 18, 2020, Jean-François Delfraissy admittedly raised a very basic question: "Perhaps we did not sufficiently appreciate the gravity of the epidemic." Jean-Paul Moatti, a former director at IRD, the French National Research Institute for Sustainable Development, confirms: "The scientific community has not been able to provide information and gain a better understanding of the situation. At the end of February/beginning of March, we could already see that South Korea's policy of combining in particular large-scale testing and contact tracing was giving positive results in containing the spread of the epidemic. With a better analysis of epidemiological curves and international comparisons, we should have been better

---

[23]Wallace-Wells D (2020, March 26) Why Was It So Hard to Raise the Alarm on the Coronavirus?, Intelligencer, https://nymag.com/intelligencer/2020/03/why-was-it-so-hard-to-raise-the-alarm-on-coronavirus.html.

[24]Juvin P (2020, April 2) « Les chefs et petits chefs de l'administration qui nous entravent face à l'urgence devront répondre de leurs actions après cette crise», Atlantico, https://www.atlantico.fr/dec ryptage/3588479/philippe-juvin--les-chefs-et-petits-chefs-de-l-administration-qui-nous-entravent-face-a-l-urgence-devront-repondre-de-leurs-actions-apres-cette-crise-coronavirus-Covid-19.

prepared to carry out mass testing.[25]" However, in its first press release on Covid-19 published as late as February 12, 2020, the French Academy of Medicine wished "to stress the importance of putting this type of epidemic into perspective. If the figures provided daily by the media may cause concern, they only express the evolution of an epidemic in a population as large and dense as that of China." In the United Kingdom, the situation was the same. Despite their prestigious institutions, the scientific and medical community was not fully aware of what was brewing and the risks posed by the new coronavirus, in part because it was convinced that any new epidemic could be handled by taking influenza as a model.

It was the same story in Belgium, a country badly hit by the pandemic. It has still one of the world's highest Covid-19 mortality (Fig. 4.1). In August 2020, three experts provided the Parliament with a first assessment of the management of the crisis. Their conclusions are clear: "How is it that we have minimised the epidemic up to this point?" Yves Coppieters, an epidemiologist at the Free University of Brussels and one of the three experts, pointed out the fact that there was "a preserve" of a few experts, which obviously "raises questions in a democratic regime" and an "inability to seek good experience elsewhere[26]". (A reaction which, by the way, is quite common: scientists are often critical about members of expert committees because they believe they should be called upon instead…). Leïla Belkhir, an infectious diseases consultant at Cliniques universitaires Saint-Luc in Brussels and another expert, explained to the parliamentarians that "Belgium had designed a pandemic plan, following the H1N1 outbreak. However, this working group was never re-established. We found ourselves having to improvise to take care of the patients."

In the *Paris-Match* magazine, on February 3, 2021, Jean-François Delfraissy responds to criticism. In the preceding days, several ministers accused the immunologist of excessive "alarmism" [at that time he advocated a third lockdown—author's note] which ruined the government's communication strategy. This is nothing particularly new; this type of confrontation occurred in September 2020 and in many other countries. After all, isn't it the role of the scientific council to be outspoken—at the risk of upsetting the government? Politicians appreciate it when scientists speak loudly

---

[25]Leglu D (2020, April 3) « Il faut renforcer l'interface science-politique», Science et Avenir, https://www.sciencesetavenir.fr/sante/jean-paul-moatti-il-faut-renforcer-l-interface-science-politique_143078.

[26]RTL Info, Les experts de la commission spéciale coronavirus sont très critiques: « Comment se fait-il que l'on ait à ce point minimisé l'épidémie?», 7 août 2020, https://www.rtl.be/info/bel gique/politique/les-experts-de-la-commission-speciale-coronavirus-sont-tres-critiques-sur-la-gestion-de-la-crise-comment-se-fait-il-que-l-on-ait-a-ce-point-minimise-l-1236028.aspx.

and publicly when they agree with government decisions. But when scientists propose an alternative, they are blamed by the politicians for stepping out of their role. However, there is also criticism from the scientific community, some doctors accusing the council of "disrupting the information." The fact remains that Jean-François Delfraissy then took on a self-imposed period of reflection and said that he would now reserve his remarks for the other members of the scientific council.

The implicit message is clear: was it necessary to create new parallel committees which complicate the work, disperse resources and blur communication lines? Are these scientific committees ultimately of any use? I cannot believe that health ministries do not have a good share of competent epidemiologists and virologists. These committees can be seen as an affront to the personnel of administrations, who are in principle competent. Not to mention the fact that these advisory boards quickly attract the wrath of those who are not nominated to be a member and who defend other scientists from the same stable... The rationale, in the end, is maybe the following: an exceptional crisis calls for exceptional measures and governments have been reassured when they have the scientific and medical elite on hand.

From the outset of the pandemic there has been a belief that scientific research is the golden thread that will lead humanity out of this nightmare. But such confidence appeared to be misplaced. Researchers have been forced to withdraw claims about once-promising drugs, the World Health Organisation has reversed key advice on combating the spread of the virus, peer-reviewed publications have been retracted after a few days, clinical trials have been short-circuited, some research projects have been supported and funded for political reasons, and the media have echoed a lot of scientific vagaries. Of course, this situation reflects the evolving state of scientific knowledge. But only in part: they are also symptoms of a deeper malaise.

The advice that we could give to future government advisers is therefore to ensure greatest transparency in the way they work and how their scientific advice that will be given to the authorities will be followed up and communicated. While they must above all ensure their independence, the advisers must also recognise that this independence will remain relative. The selected experts are appointed by the government but they are, for most of them, paid by one or several employers, public and/or private, for a range of contributions. Who can really claim, under these conditions, to be totally free and to provide "independent" opinions in a "scientific" committee? There are potentially many conflicts of interest. Yet expressions such as "independent audit" and "independent expert committee" are legion! The myth of independence does persist. And here I do not mention the pressures from the authorities and the stress created by constant exposure in the media. We have seen many examples of this in the previous pages: the ruling power will always try to control, by various means, direct and indirect, the contribution and the communication of the experts. And if things are getting worse, they will marginalise or even remove them.

The experts must also give credit where credit is due. Science progresses through multiple, rigorous and somewhat lengthy experiments and verifications. Letting the idea spread that a vaccine can be obtained in a few weeks or that the effectiveness of a drug can be demonstrated in a few days is not only counterproductive but dangerous. There has been, in some cases, a serious breach of the ethical rules in the design of the research protocols and the communication of their results, thus opening the door to uncertain conclusions and fake news. The objective to advance as quickly as possible, presented as a demand from society, leads to the acceleration of poor science, bypassing peer review processes, encouraging premature announcements of preliminary results and, given the financial stake, giving rise to organised disinformation campaigns. The worst outcome is when political leadership comes to support and even participate in these manipulations, like Donald Trump and Jair Bolsonaro, who both bear a personal responsibility in the evolution of the epidemic in their respective countries.

Let us also remember that science and experts are not the absolute guarantee of which political leaders dream. If no one disputes the important and positive role played by Sotirios Tsiodras in Greece, Christian Drosten in Germany and Anthony Fauci in the United States, the contribution of Anders Tegnell in Sweden is very controversial, where the ideas of the epidemiologist in favour of herd immunity are no longer as seductive as they once were. It is even worse in France where the professional misconduct by Didier Raoult led the government to authorise the prescription of hydroxychloroquine for a few weeks, in the absence of scientific work worthy of the name, and then to abolish it. And in the United Kingdom, scientific and medical experts advising the government ignored signals from China and Italy, and were spreading the idea that the new epidemic would be handled as a seasonal flu. Two months, February and March 2020, were lost, time in which the country could have been preparing to face the virus.

Science is advancing through the study of simple systems, whose various parameters can be controlled and the observed behaviour easily modelled. But the "scaling-up" to real life most often requires a qualitative change. These are systems in which many additional factors come into play, some known, others unknown, and taking into account the fact that some experimental data may be unavailable or is lacking precision. Modelling an epidemic is possible from a scientific point of view, but there are limits when it comes to simulating "live" the course of a new disease in real time. Scientists did their best to follow the ongoing pandemic, with incomplete and sometimes erroneous data. This is a real scientific issue, with of course important health issues and methodological questions at stake. But, as a group of researchers and physicians reminds us, "scientists are free to use their own hypotheses and methods to defend ideas which may go against the scientific grain. But to convince their peers, they must provide transparent, comprehensive and reproducible results, so that the scientific community can verify their claims—an essential prerequisite to providing a rapid benefit to mankind. In science, credibility is built on solid discovery and robust evidence, not the other way round.[27]"

While it is accepted that scientists advise decision-makers, it is less often recognised that politicians can also influence scientists. The issue of face masks is a good illustration of this. In Europe and the United States, wearing a mask was not recommended for the population until the beginning of April 2020. This was also WHO's position. Then most governments made a U-turn. Some scientists followed. Like the prestigious Royal Society, which published at the beginning of May a report supporting the use of a mask by the general public.[28] However, these conclusions fuelled a fire of criticism from several experts, considering in particular that there was still, at that time, no scientific proof that face masks reduce the transmission of viruses. Does this mean that scientists can also bend their conclusions to the needs of politicians? Or is it because the knowledge was there and the publication considered as timely? Remember also the changes made by France's scientific council on the question of schools, first proposing on April 20, 2020 "to keep nurseries, schools, colleges, high schools and universities closed until September 2020," and then, four days later, accepting "the political decision" to reopen schools from May 11."

When I hear experts in virology and doctors of epidemiology talking on prime-time television or radio, they frankly sound like politicians. Their self-confidence can pass for arrogance and some, in the end, preach for their own chapel: here you have the lockdown ideologists, there the modelling partisans, the vaccination activists, the "testology" supporters, etc. In short, they all have a line—or even a party line—to defend… I noticed that some scientists who are used to appearing on prime time and mainstream media tend to

[27] Collectif de chercheurs et de professionnels médicaux (2020, September 2) Halte à la fraude scientifique, Libération, https://www.liberation.fr/debats/2020/09/02/halte-a-la-fraude-scientifique_1798277.

[28] Royal Society DELVE Initiative (Data Evaluation and Learning for Viral Epidemics), Face Masks for the General Public (2020, May 4) https://rs-delve.github.io/reports/2020/05/04/face-masks-for-the-general-public.html.

become less scientific… Science is universal, but when it diffuses into society, it goes through filters and can get slowed down or accelerated. Like any human being, the scientist also appreciates seeing his or her ideas accepted by society. It is not a question here of the theories or models that he or she develops—which are ultimately very "technical" and are not, moreover, open to "discussion"—but of the hypotheses on which his or her approach is based or of the interpretations to which his or her observations or calculations can lead.

In any case, the handling of the Covid-19 epidemic in France, the United Kingdom and the United States shows that the cooperation between science and politics has its limits. Emmanuel Macron and Boris Johnson repeatedly reiterated their confidence in science or, to be more precise, in "their" scientists, who were supposed to represent the whole scientific community. Donald Trump, although he never showed he was relying on science, nevertheless wanted his scientific advisers to accompany him (but stay in the background). However, they all, at some point, kept their distance from science. This is understandable: a political decision is not a scientific statement. In the process when leading a government, there is always a point where scientists will lose ground. Conversely, the rationale put forward by scientists often does not allow politicians to *reconnect* with the real world. This is what the German sociologist Max Weber used to say more than a century ago, emphasising this fundamental distinction between them both: a politician takes a stand, a scientist analyses the world. According to Weber, scientists should not get involved in politics and must preserve their authority, which guarantees the understanding of facts. Claiming that politics can follow science is a deception. In these three countries, the problem was not tackled correctly from the start.

However, this is not all we have seen. The Covid-19 crisis actually stimulated the emergence of "science politics"—politics by science. The pandemic has prompted new scientific controversies, widely disseminated by the media, about how to cure people affected by the coronavirus. These have been highly politicised. Many experts across the world have used the media and their role as advisers to promote their own views, non-scientific opinions, ideological positions and decisions for society—including on issues unrelated with the pandemic—therefore entering the field of politics. The emblematic case here is obviously Didier Raoult. When he criticised the media for their incompetence and politicians for creating a climate of fear, he certainly left the field of science to move into "scientific populism." Promoting herd immunity is a political choice. More worryingly, Neil Ferguson in the UK, Anders Tegnell is Sweden, and Deborah Birx in the US made errors revealing their disrespect for their own professional commitments. The phenomenon is not new but it developed at an unprecedented rate. The "politics" of science became front-of-stage in these conflicts, unveiling links between the production of science and the private sector, personal struggles among researchers who strive for recognition and authority, and the intermingling of science with politics. For sure, the pandemic provided these experts with huge visibility in the media and a captive audience across the world. According to public opinion surveys, they had even become among the most popular "political personalities" in their countries. And if we have seen them fighting with politicians, it is also because they were stepping on the toes of their government 'masters' and aiming to play at the same level. Conversely, some experts remained silent because they did not want to oppose the government and put themselves in a difficult situation. In either way, they were all playing politics. By science politics, I mean highlighting research works, real or potential, which go in the direction desired by the authorities. Many examples illustrate this in the previous pages. But I also want to address the fact that much of the scientific community holds the idea that research is the solution to our societal problems—from health and energy to the environment and communication. This ideology of progress, actively supported by researchers, provides considerable benefits in terms of funding. By advising governments, experts adhere, at the very least explicitly, to the old myth that science can rule the world. Could science become a political party?

Which "Public" Service?

The health sector is becoming increasingly privatised across Europe. This is a marked trend, which transcends political divisions. More and more, the public hospital is seen as a "centre of profitability" and is driven as a commercial enterprise. Hospital managers are requested to project themselves into the globalised healthcare "market" and to demonstrate that they are more efficient than their "competitors." May be one day we will say this was an error—or a fault. It is anyway part of the general evolution of society. Even Germany, where the hospital system is particularly robust, was under severe strain during the crisis. Although the course of the epidemic there has been kept under control, the country had to face a shortage of medical and nursing staff, bemoaned by health professionals for years. Workload, reduction of investments and poor working conditions are often blamed by healthcare workers as the consequences of a system mainly focused on cost–benefit considerations.

Can we privatise public services? We answer the question by asking it. The Covid-19 crisis has highlighted the problems raised by bringing to market those sectors of the economy that are primarily services to society. Of course, privatising is a comfortable solution for our leaders since it provides them with cashflow and avoids having to find elaborately complex and long-term solutions. This is not a priori incompatible with the notion of public service. However, generalised competition and selecting the lowest tender instead of the best offer change the rules of the game. Will the average patient still have a bed in the hospital? How can we guarantee access to adequate health care for everybody, knowing also that health systems that resist privatisation, such as the NHS in the UK, do not always perform any better?

These legitimate questions are certainly not new. In some sectors, such as transport, privatisation is no longer on top of the political agenda. But it is also upstream that the alerts from the scientific and medical community have been overlooked. Coronavirus scientists have repeatedly denounced the lack of long-term funding to support the development of broad-spectrum antiviral molecules. More fundamentally, the scientific community has been urging governments for years to provide more resources for research. We cannot praise science and technology and at the same time neglect basic research. In Europe, this fight has become a litany of good words. In contrast to China, which is experiencing the largest increase in R&D budget in the world. The Covid-19 crisis has clearly shown that there is a heavy price to pay for modest levels of research expenditures.[29]

These questions are more legitimate than ever because a large fraction of the population feels that, in our societies, technological progress and democratic control have diverged. The question here is not only the role played by science in the development of military technologies and totalitarianisms in the twentieth century,[30] but the proliferation of machines that we have not really asked for and plentiful technologies which are at best marginal (the connected watch) or controversial (cloning) and at worst counterproductive (the electric car) or dangerous (nuclear weapons). All in all, our society looks quite schizophrenic as we continue to devote scandalous resources to the development of gadgets and superfluous innovations while, on the other hand, calling for more action on problems that will be the key to

---

[29]In 2018, the R&D expenditure (as a percentage of GDP), stood at 2.18% in the EU, much lower than in South Korea (4.53%), Japan (3.28%) and the United States (2.82%) and about the same level as in China (2.14%): https://ec.europa.eu/eurostat/statistics-explained/index.php/R_%26_D_expenditure.

[30]The reference here is Benno Müller-Hill, who has shown that German scientists actively contributed to the implementation of the Nazi ideology.

the future of humanity (climate change, "green" agriculture, clean energies, cancer treatment, etc.).

It is often said that all it takes is a crisis to unlock budgets. This is sadly true. What we are experiencing today confirms it once again: following an unexpected pandemic, politicians have woken up and funds are pouring in. Then we all remember what we should never forget: research is a very, very long-term investment but it always pays-off. The problem: we don't know when and we don't know what. In the short-term, some research may look unsuccessful, because of epistemological bottlenecks or lack of funding. In science, we do not always carry out research where we want to but where we are able to. The vaccine developed by Pfizer and BioNTech was a good reminder of this: the concept of a messenger RNA vaccine had been explored for several decades but, due to lack of resources, it could not have been validated. The budgets released in the emergence of the crisis provided the necessary boost.

It is unfortunate, however, that it takes a major crisis for the public to realise the vital role of many professions, and not just in the health and science fields, but also in the general services, social care and many others. Likewise, governments should have realised that the health sector and pharmaceuticals are not competitive markets, in particular due to the high number of monopolies. Moreover, some markets do not work in times of a pandemic, such as essential supplies (masks, drugs, reanimators, etc.). If our fight against epidemics is based on a purely commercial approach, it will cause millions of deaths in developing countries and even more victims in developed countries.

This crisis, initiated by a viral epidemic, is in reality a global, systemic crisis, which has been amplified by factors external to health—economic, ecological, political, financial, structural. It is the product of neoliberalism, incompetence and widespread indebtedness. It will probably have—nothing is more certain in fact—far-reaching consequences. In a world that over-values athletes, media stars and financiers, what place do we want to leave to researchers? To caregivers? To rubbish collectors? To educators? To peace-keepers? The crisis has shed light on those who work in the shadows, those who care for the sick and help the vulnerable. Which vaccine are we going to develop to protect them?

Faced with a crisis or a catastrophe, a natural reaction is to turn to political authorities because these events require a massive and collective reaction. For Nobel laureate in economics Joseph Stiglitz, there is a positive side to the current crisis. Paraphrasing the popular expression, he puts the church back in the middle of village life: the current crisis highlights the fact that health should be considered a fundamental and universal right at all levels of society.[31] In Europe, this vision gave rise to what has been called the "welfare state." It is at the heart of the European social model whereby health, education, universities, pensions, social security and other public services are financed mainly by public action.

Maybe this social model will emerge from the crisis. I often have discussions with my American, Chinese, Korean, Indian, Japanese and Russian colleagues about the quality of life in our respective countries. Europe is always seen as a benchmark. Of course, taxes are high, but in return, the essentials are guaranteed, and for life. The net wages are not very different in Europe and in the United States but on the Old Continent, almost everyone benefits from medical insurance, access to affordable education, social security, etc. The big difference is over the long term. In Europe, we are in principle protected against the injustices of life (disease, job loss) and we can offer a university education to our children even if our income is modest. Compared to the United States and China among others, high-level education here is almost free (except in the UK) and most students graduate without having a debt they will carry for the rest of their life (many Americans are forced to postpone their retirement because their student loans are not yet repaid). Vigilance is required, however, as our governments actively watch for any possibility for further privatisation and increased competition.

This type of discussion comes up with every crisis or major accident. Except that in this case, almost every country in the world has been hit. Are we collectively going to rethink our priorities? Or will we keep our head in the sand? Either way, the world will not be the way it used to be.

Fighting the Infodemic

Why has the Covid-19 viral pandemic become an economic, diplomatic, political and let's face it global crisis? Without a doubt, the novelty of the virus, the speed of the epidemic's spread and the number of countries affected have turned our societies upside down. But have we also given in to the pressure of the media? The answer is yes.

This pandemic was marked by a change in scientific communication. The visibility given to research still in progress or even at an early stage

---

[31] Stiglitz JE (2020, March 13) Plagued by Trumpism, Social Europe, https://www.socialeurope.eu/plagued-by-trumpism.

of discussion (and also to pseudoscience) has been disproportionate.[32] We have followed and endured the manoeuvres of some scientists and companies to promote their vaccines, treatments and other remedies, a sign that commercial developments are at stake. This crisis has also revealed the huge credibility now given to "fake news" and conspiracy theories. Anguished and disoriented, the public no longer knows which way to turn and conspiracy theories were most prominent at the height of the pandemic, when anxiety and uncertainty are high, as shown by an international team of scientists.[33] Communicators (the media), just as much as the non-communicators (governments), could not escape criticism. There has also been a fair amount of frustration aimed at social media platforms who are accused of spreading this fake news. However, the real influence of the "new media" does not seem to be very significant: according to an opinion poll carried out in Sweden between March 18 and March 21, 2020, people who obtained information on the coronavirus from social networks account for only two percent.[34] Researchers have shown that traditional media outlets remain a major source of information for many people due to their perceived credibility during a crisis, as observed in particular in Italy[35] and France.[36] Nevertheless: there has been so much misinformation and disinformation flying around (the latter being different from the former by a desire to deceive) that WHO set up a new unit and a specific strategy dedicated to counteracting the negative effects of the *infodemic*, defined as "an overabundance of information—some accurate and some not—that makes it hard for people to find trustworthy sources and reliable guidance when they need it.[37]" The examples of mis- and disinformation run into the thousands and touch on all areas. I just mention a few here[38]: "Pet animals are the sources of coronavirus"; "Eating garlic

---

[32]Scheirer W (2020, July 20) A pandemic of bad science, Bulletin of the Atomic Scientists, Volume 76, https://www.tandfonline.com/doi/full/10.1080/00963402.2020.1778361.

[33]J. Metcalfe et al., The Covid-19 mirror: reflecting science-society relationships across 11 countries, JCOM 19 (07), A05, https://doi.org/10.22323/2.19070205.

[34]https://v-a.se/2020/04/coronavirus-in-the-swedish-media-study-high-public-confidence-in-resear chers-and-healthcare-professionals/.

[35]M. Bucchi, N. Saracino, Italian citizens and Covid-19: one month later — April 2020, *Public Understanding of Science Blog*, April 19, 2020, https://sagepus.blogspot.com/2020/04/italian-citizens-and-Covid-19-one-month.html.

[36]Carasco A (2021, January 26), La crise du Covid-19 réconcilie (un peu) les Français et les médias, La Croix, https://www.la-croix.com/Economie/crise-Covid-19-reconcilie-peu-Francais-medias-2021-01-26-1201137240.

[37]WHO (2020, February 2), Novel Coronavirus (2019-nCoV) Situation Report – 13, https://www.who.int/docs/default-source/coronaviruse/situation-reports/20200202-sitrep-13-ncov-v3.pdf.

[38]Islam MD et al., (2020 October 7) Covid-19–Related Infodemic and Its Impact on Public Health: A Global Social Media Analysis, The American Journal of Tropical Medicine and Hygiene, https://doi.org/10.4269/ajtmh.20-0812.

can cure coronavirus"; "Cannabis boosts immunity against the novel coronavirus"; "COVID-2019 outbreak was planned"; "President Donald Trump targeted the city with coronavirus to damage its culture and honour in Iran"; "Every disease ever has come from China"; "Bill Gates is in possession of inside information about Covid-19."

Why has this viral epidemic been accompanied by a media pandemic? It is not the first time that confidence in science and medicine has taken a hit by the proliferation of false information on the internet. In some ways fear was fuelled by the experts, who recognised in the earliest stages of the outbreak that science could offer no help—or almost no help—against the coronavirus. This was just too hard to believe: a hyper-technological society like ours had no other solution than to resort to the medieval practice of quarantine!

Science and medicine contribute to our dreams and fears in our perception of the world. This is why technological accidents and health crises always generate a lot of media coverage. Deluged with numbers, statistics (which largely emphasised the steady increase in the number of cases), looped images and statements, the public lost sight of the fact that mortality remains low (less than one percent) and that the lethality is apparently decreasing.[39] As seen with previous crises and accidents, there is no direct link between the extent of media coverage and the death toll. Let us just recall the huge influence the Fukushima Daiichi accident had on us compared to its real impact—zero deaths!

A brief reminder: what we know about Covid-19 and the coronavirus does not come directly to us from the laboratories but from what I call "mediascience", that is science that we receive through the media.[40]

Mediascience is not real science: it is filtered by news media, digital platforms, social media and video sharing sites, and served to us through different channels, often repackaged according to the latest fashion of communication. But mediascience is a special kind of science, both in substance and in form. It shines a light on specific (sometimes anecdotal) research but keeps the bulk of it in the dark! Mediascience is also about images, people and stories, that may influence the public. In short, mediascience is to science and medicine (which is seen by Europeans as "the most scientific of sciences"[41]) what the news is to general information. Eurobarometers, the opinion polls carried out in European Union Member States, have shown

---

[39]Ledford H (2020, November 11) Why do COVID death rates seem to be falling?, Nature, https://www.nature.com/articles/d41586-020-03132-4.

[40]Claessens M (2011) Allo la science ?, Paris, Hermann.

[41]European Commission (2001), Europeans, science and technology, https://ec.europa.eu/commfront office/publicopinion/archives/ebs/ebs_154_en.pdf.

that technological accidents and health crises generate a great deal of scientific information which is disseminated to the public. For better or worse, mediascience contributes to knowledge dissemination, to public understanding and even to the advancement of science! A somewhat anecdotal—but significant—example: converging arguments support the fact that at the end of the last century the Hollywood production Jurassic Park increased the level of knowledge of dinosaurs for the general public which directly led to many a choice of vocation among young people…

In a crisis such as Covid-19, mediascience mainly reports on *research* being done and not about accomplished *science*. This is quite a fundamental difference. Indeed, citizens expect scientists to have the answers to their questions and problems, but research does not provide them immediately. This is evidenced by the many discrepancies, unverified interpretations and contradictory opinions between researchers and experts that have been exposed in the media. Science is a truth which is built over the long-term, anthropologist Gilles Boëtsch recalls.[42]

This brings us to this central question: should we be afraid of the coronavirus? Based on current data, the answer is no—or should be no. While the risk of infection and death is not zero, it is still low and much lower than other agents and fatal causes: 98% of people infected by the coronavirus recover. The total death toll from Covid-19 is currently more than 10 million while each year 18 million people die of cardiovascular disease and nearly 3 million children die from malnutrition. However, compared to the coronavirus, these deaths no longer count. They even no longer *exist*. In this new digital communication age, we tend to live in the media sphere, and hence only see what circulates on the networks. If your death is not announced on Facebook or Twitter, it doesn't mean you are still alive, but simply that you don't really *exist*.

This very close relationship between mediascience news and the perception of a technoscientific risk is very well illustrated by Covid-19. It also explains why close monitoring of the media is used to detect early warnings of a crisis or a controversy and to ensure an appropriate follow-up even if everybody agrees that there is, as we say, a non-negligible "background noise."

When it comes to risk and, more specifically, to the public's perception of risk, mediascience is inevitably forging our understanding with non-scientific arguments. It is quite a challenging job to talk about science when there are so many catastrophic scenarios and shocking images that are

---

[42]Stenvot L and Testard-Vaillant P (2021, March), Les premières leçons de la crise du Covid-19, CNRS Le Journal, p. 30–31, https://lejournal.cnrs.fr/sites/default/files/numeros_papier/jdc303_total_web.pdf.

around us everywhere and every day. Mediascience is a distorting—and also a distorted—mirror of science, and our appreciation of risk is influenced if not determined by irrational factors, such as the diffuse feeling, confirmed by sociologists, to live in a "society of fear." This is in turn amplified by the media because it is part of their core business and for reasons which are more of a commercial nature. And also, because journalists, reporters, bloggers and influencers are human beings! Our fears feed their fears, and then the media, which feeds them in return. Thus, a good example of a vicious circle. This is particularly true in the field of health, where our powerful and omnipresent technology has paradoxically nurtured a permanent anxiety, which is linked at least in part to the sophistication of the instruments used by physicians and care workers. In short, our health is now like our car: we cannot repair it alone! We depend on various people and instruments—in fact a whole system. Understanding the risks concerning our health—which also include the risks of possible treatment—requires medical staff to communicate about the uncertainties and potential side effects, which in turn reinforces our anxiety.

Scientists love to say that science journalists so enjoy controversies that sometimes it is they who are driving researchers and are chasing after scientific disputes, which allow them to report on the shortcomings of theories and models and their conflicting interpretations. Actually, it is mostly the other way around. It is when the media is focusing on breaking news like crises, discoveries and breakthroughs that journalists and others in the media get to learn about the different and sometimes divergent approaches of the specialists in the field. In their turn, they will report and echo the coexistence of several theories and therefore bring the (still) obscure side of science into the spotlight. It is quite healthy in my opinion, and in any case is a fundamental aspect of the scientific process. And it is also the very nature of the journalistic method. This explains why Covid-19 is, from a mediascience point of view, an amazing subject. It concerns science, people and the interactions between science and society. Although the public might not be ready to hear all that. It is as much the tragic dimension of the pandemic and the scientific community's efforts to find the solution to this crisis that both terrifies and fascinates people at the same time—fears and hopes.

One good thing is that this visibility also impacts on the research community, as scientists are also watching with rapt attention. Some of them are even devoting more time and effort to mediascience than primary science. Unsurprisingly, the media are more than just observers: they also play a role in these crises and participate in the emergence, development and death of scientific controversies. We have seen this in the case of hydroxychloroquine, where public exposure of the key proponents allowed ongoing research to speed up and improve while unfortunately precipitating wrong decisions made by some governments. Media science is indeed a reality—for better or worse.

But let's be clear: mediascientific choices are ultimately editorial decisions—therefore not very transparent—which may follow personal or organisational motivations. Having organised dozens of press conferences in my career, I was sometimes frustrated by the fact that the public visibility given to a particular research project actually prevented other research from being seen and at the same time attracted a lot of opposition, incomprehension and even jealousy from those researchers whose work remained in the shadow. There is also the risk of placing too much importance on "mediascientific" researchers, those who are at ease with public communication and master the rules of the media game.

Hydroxychloroquine was a mediascientific topic before being a scientific issue. The drug got promoted, among many others, by Didier Raoult in France and Donald Trump in the United States. However, several works, published in reputable medical journals, reported and confirmed negative results. A (now retracted) study published on May 22, 2020 in *The Lancet* and conducted by Mandeep R. Mehra (Harvard Medical School) was seen as dampening hopes.[43] This international retrospective registry analysis, the largest to date, reviewed the medical records of some 96,032 hospitalised patients, of whom 14,888 received either chloroquine or hydroxychloroquine, sometimes in combination with an antibiotic, and concluded that neither of these drugs were effective against Covid-19. On the contrary, these molecules turned out to increase the risk of death and cardiac arrhythmia and the authors of the study recommend against prescribing them outside of clinical trials. The authors estimate that the risk of mortality is 34% to 45% higher in patients taking these treatments than in patients with comorbidity factors. The same day, Olivier Véran asks the HCSP to propose a revision "within 48 h" of the derogatory provisions authorising hydroxychloroquine-based treatments in France. Then, proponents rightly argue that Professor Raoult advocates the administration of hydroxychloroquine in combination with azithromycin, which does not correspond with the protocol reported in *The Lancet*. However, on May 26, the HCSP and the National Medicines Safety Agency (ANSM) decide to suspend the authorisation to prescribe hydroxychloroquine to hospitalised Covid-19 patients. France is followed by

[43] Mehra MR, Desai SS, Ruschitzka F and Patel AN (2020, May 22) Hydroxychloroquine or chloroquine with or without a macrolide for treatment of Covid-19: a multinational registry analysis, The Lancet, https://doi.org/10.1016/S0140-6736(20)31180-6.

WHO, which withdraws hydroxychloroquine from its international "Solidarity" clinical trial conducted to test four potential treatments against SARS-CoV-2: remdesivir, lopinavir/ritonavir, interferon-1a and hydroxychloroquine.

However, the controversy is not yet over. Didier Raoult describes the publication in *The Lancet* as "messy" and in a tweet points to the homogeneity of the profiles of the patients selected for the study, even though they come from five continents. He also underlines the fact that the number of Australian Covid-19 deaths given in the article does not match the data from the authorities in Canberra. Didier Raoult also expresses his surprise that the study data had been collected by a small company, Surgisphere, situated near Chicago. As a result, several researchers request to have access to all the experimental data in order to verify how they were collected and analysed. *The Lancet* acknowledges an encoding error (an Asian hospital having been registered as Australian) but confirms the validity of the findings. Only a large-scale randomised study could probably shed light on this issue. Some scientists argue that discussing complex subjects like this one in the public square does not help anyone. The technical issues are in fact not debatable—at least with the public. This is why scientists can seldom offer definitive answers in a public debate. Some issues are too complex to fit within the current scope of science, or there may be little reliable information available, or the values involved may lie outside of science, and very often there is just not enough time allowed for a detailed explanation. Public debates often concentrate on the non-scientific part of the work and the most controversial issues. This is also the very nature of science: philosophers of science have shown that controversies are inherent in research. Science even needs them to move forward and make progress.[44]

But the case reappears a few days later. Several readers contact *The Lancet* and ask questions about the demographic data of the study and the doses of hydroxychloroquine used. On June 2, the medical journal publishes an "Expression Of Concern" (EOC) about the data in the article and notes that an independent audit of the provenance and validity of the data has been commissioned by [some of] the authors. A few hours earlier, it was another prestigious journal, the *New England Journal of Medicine*, that publishes an EOC about one of its articles, published on May 1, 2020. The publication, the data of which was also provided by Surgisphere, claims that taking certain blood pressure drugs like angiotensin-converting enzyme (ACE) inhibitors does not increase the risk of death among Covid-19 patients. And there is

---

[44]Lecompte F (2020, April 6) Edgar Morin: « Nous devons vivre avec l'incertitude», CNRS Le Journal, https://lejournal.cnrs.fr/articles/edgar-morin-nous-devons-vivre-avec-lincertitude.

even a third article, in preprint since April 2020, that is being questioned because one of the authors is none other than the founder of Surgisphere, Sapan Desai. This article (now also retracted from the site that hosted it) reportedly concluded that ivermectin, an antiparasitic drug, drastically reduces the mortality of Covid-19 patients. On June 4, 2020, three of the four authors of the *Lancet* article, Mandeep R. Mehra, Amit Patel and Frank Ruschitzka, request, without Sapan Desai, "that [their] paper be retracted." They add that "our reviewers were not able to conduct an independent and private peer review and therefore notified us of their withdrawal from the peer-review process.[45]" Surgisphere apparently refused to disclose the full data set used for the study. The study's lead author, Mandeep R. Mehra, is the only person to shows contrition, specifically in an official press release published on the website of the Brigham and Women's Hospital: "I have always performed my research in accordance with the highest ethical and professional guidelines. However, we can never forget the responsibility we have as researchers to scrupulously ensure that we rely on data sources that adhere to our high standards. It is now clear to me that in my hope to contribute to this research during a time of great need, I did not do enough to ensure that the data source was appropriate for this use. For that, and for all the disruptions—both directly and indirectly—I am truly sorry.[46]" For its part, *the New England Journal of Medicine* also announced the retraction of the article published on May 1. In both publications, the lead author was Mandeep R. Mehra, a professor at Harvard Medical School and a director of the Brigham and Women's Hospital Heart and Vascular Center in Boston, Massachusetts.

In short, it is the scientific integrity of the researchers and contractors involved that is put into question by this extraordinary affair. Another surprise comes from the fact that WHO resumes their clinical trials of hydroxychloroquine on June 3. A carefully-chosen date or a calendar coincidence? After analysis of the available data, the Director-General, Tedros Adhanom Ghebreyesus, specifies during a press conference that day "that there are no reasons to modify the trial protocol" of Solidarity, and that the Data Safety and Monitoring Committee has not detected any increased mortality due to hydroxychloroquine at this stage. WHO scientist Soumya

---

[45]Mehra MR, Ruschitzka F and Patel AN (2020, June 4) Retraction—Hydroxychloroquine or chloroquine with or without a macrolide for treatment of Covid-19: a multinational registry analysis, The Lancet, https://www.thelancet.com/lancet/article/s0140673620313246?utm_campaign=tlcoronavirus20&utm_source=twitter&utm_medium=social.

[46]Mehra M. R. (2020, May 22) No Improvement in Death Rate for Covid-19 Patients who Received Hydroxychloroquine, press release of the Brigham and Women's Hospital, https://www.brighamandwomens.org/about-bwh/newsroom/press-releases-detail?id=3592.

Swaminathan, for her part, says that "there are so many potential biases in the way that patients are managed in regular clinical setting, that the only way to get definitive answers is to do well conducted randomised trials [where the patients are chosen at random, as well as the treatment which they receive— author's note]. And it's particularly important in emergency settings to do this because that's the only way to find out what really are those drugs or those strategies that will reduce illness, that will reduce infection rates in communities. And we should be guided by the science and by the evidence.[47]" Still, it is tempting to see WHO's decision as a U-turn about-face resulting from political and media pressures on the organisation.

This sequence of events and announcements, led by two of the most prestigious medical journals, was obviously not met by indifference by the scientific community. Carlos Chaccour, a physician working at the research institute ISGlobal and a renowned expert in ivermectin and parasitic infections, summarises for the newspaper *The Guardian* what the majority of scientists are thinking. He was initially intrigued by the study, which was published on 14 April 2020 in a preprint, and impressed by the fact that the authors were able to access a huge database, which included 169 hospitals in Asia, Europe, Africa, North America and South America. Yet, Chaccour soon found some concerning anomalies. Those doubts would only increase over the next weeks. Surgisphere itself came under greater scrutiny, culminating in two of the world's most prestigious medical journals withdrawing two publications based on its data. The scientist concludes: "We are facing a pandemic that has claimed hundreds of thousands of lives, and the two most prestigious medical journals are not doing their job."[48] The case gives an idea of the challenges scientific research is facing when it has to work under intense pressure and balance private interests.

One sign that science is under pressure is flagged up in the United Kingdom. On June 3, 2020, the UK's Medicines and Healthcare products Regulatory Agency (MHRA) asks the independent committee evaluating the Recovery trial to look at the latest data. Recovery is also a randomised trial, which enrolled over 11 000 patients in 175 NHS hospitals. On June 5, the two coordinators of the trial, Peter Horby and Martin Landray of the University of Oxford unveil the conclusions of the evaluators: "Hydroxychloroquine does not work against Covid-19 and should not be given to any more hospital

[47]WHO (2020, June 3) World Health Organization Coronavirus Press Conference June 3, https://www.rev.com/blog/transcripts/world-health-organization-who-coronavirus-press-conference-june-3.

[48]Davey M (2020, June 4) Unreliable data: how doubt snowballed over Covid-19 drug research that swept the world, The Guardian, https://www.theguardian.com/world/2020/jun/04/unreliable-data-doubt-snowballed-Covid-19-drug-research-surgisphere-coronavirus-hydroxychloroquine.

patients around the world.[49]" But is this study correct? Martin Landray says the hype should stop now: "Hydroxychloroquine is being touted as a game-changer, a wonderful drug, a breakthrough. However, today we can stop using a drug that is useless." In fact, a WHO expert panel had already strongly advised against use of the anti-inflammatory drug to prevent infection in people who do not have Covid-19 on March 2, 2021. Their recommendation was based on six randomised controlled trials involving over 6 000 participants with and without known exposure to a person with Covid-19 infection.[50]

Nevertheless, these results did not prevent Donald Trump, Jair Bolsonaro and others from continuing to take hydroxychloroquine regularly and speaking about it publicly. Once again, top officials forget that leadership means leading by example. But this series of events is not yet finished: on June 16, 2020, the Recovery Collaborative Group announces that dexamethasone, a corticosteroid drug known for its anti-inflammatory effects, reduces mortality in the most seriously ill patients by one third, a discovery immediately described by WHO as a "scientific breakthrough."

At a hearing on June 24, 2020 before the Parliamentary Commission of Inquiry of the National Assembly on the management of the Covid-19 crisis, Didier Raoult blasts the government, the ANSM, HAS as well as the editorial staff of The Lancet. Professor Raoult appears as a scientific version of Donald Trump. Unsurprisingly, he becomes the target of fierce criticisms from several of his colleagues for his "slander" (HCSP), "infamous and unfounded remarks" (scientific council), and of having issued "false testimony" (Public Assistance - Paris Hospitals). He also has to face a complaint lodged in July 2020 by the French Society of Infectious Pathology, which considers that the doctor of Marseille has violated nine articles of the code of ethics of the profession, in particular by promoting a treatment whose effectiveness has not been demonstrated, by disseminating false information, by breaching the duty of confraternity and by carrying out clinical trials at the limit of legality. The association also criticises Didier Raoult for having repeatedly announced the end of the epidemic and for having called those who does not prescribe hydroxychloroquine "crazy". It is on this basis that, on November 9, 2020, Professor Raoult was informed of a complaint filed by the Medical Order of the department of Bouches-du-Rhône against him, in particular for violation

---

[49]Boseley S (2020, June 5) Hydroxychloroquine does not cure Covid-19, say drug trial chiefs, The Guardian, https://www.theguardian.com/world/2020/jun/05/hydroxychloroquine-does-not-cure-Covid-19-say-drug-trial-chiefs.
[50]Lamontagne F et al. (2021, March 2) A living WHO guideline on drugs to prevent Covid-19, BMJ 2021; 372, https://doi.org/10.1136/bmj.n526.

of confraternity, false information to the public, exposure to undue risk and charlatanism for the promotion of hydroxychloroquine.

It is ultimately a seemingly scientific debate that sheds light on professional interests —combined with personal rivalries—as well as industrial issues underlying the coronavirus research, of which some players are using ideological arguments to try and consolidate their position. Usually, it is quite the opposite: science is often used to support ideology. Some scientists are keen to talk to the media to mark their territory and publicise their personal thoughts. In times of crisis, some scientists are pushing the limits of science, others are using the uncertainties of research to their own advantage. All this reminds us that the border between science and politics is very porous. How can it be otherwise? We have to think about "science in society" as soon as scientists leave the laboratory and become involved in major economic, political and societal issues. To conclude the hydroxychloroquine case, I am surprised, as a scientist, that a scientific publication can be accepted by the most prestigious journals while its data has been collected by a subcontractor, whose manager is unable to communicate the set to his co-authors! Any researcher is well aware that he or she must keep control of the experimental data, know precisely how it is measured and collected in order to describe in details, for example in a subsequent article, the experimental protocol. When I was a doctoral student, it was almost unimaginable that one could "privatise" the experimental part of research work.

The "Lancetgate" scandal also shed light on the very profitable business of medical data, as some companies succeed in "siphoning off" medical records by hacking or via imaging systems located in hospitals—much like some of the digital giants have done in the past (which led to widespread condemnation for personal data breaches and abuse of a dominant position). In such a context, where scientific, industrial and political interests mix, the impact of mediascience, far from being negligible, nevertheless remains limited: science is not the antidote to fake news and conspiracy theories.

At the same time, another controversy is rumbling about the vaccine candidates developed by nearly two hundred laboratories: a frenzied and merciless race which is fuelled by the health crisis and the formidable financial stakes involved. In this case too, some players are bypassing the usual procedures in order to reduce the production times from several years to a few months. When Vladimir Putin announced on August 11, 2020 that the first vaccine had been developed by the Gamaleya Research Institute of Epidemiology and Microbiology, experts pointed out that the Moscow laboratory broke with usual scientific protocols to speed up the production process. Of course,

Table 4.2 Production milestones of the Moderna's mRNA vaccine

December 1, 2019	Covid-19 documented
January 11, 2020	SARS-CoV-2 virus sequenced
January 16, 2020	NIH designs mRNA vaccine with Moderna
March 16, 2020	Clinical trial phase I/II begins (to assess safety, side effects, best dose)
July 27, 2020	Clinical trial phase III trial begins (to determine therapeutic effect)
October 22, 2020	Enrolment in phase III complete, > 74 000 participants
November 16, 2020	Moderna announces 94.5% interim efficacy
December 11, 2020	FDA issues EUA for Pfizer/BioNTech vaccine
December 11, 2020	Vaccination begins for health care professionals
December 18, 2020	FDA issues EUA for Moderna vaccine

Russia has a regulatory procedure that allows, in the event of a health emergency, the authorisation of a vaccine that does not yet have full approval. French expert Marie-Paule Kieny believes that all stages of development have still been respected despite the short deadlines: "Although usually these stages are successive, they follow on very quickly in the case of Covid-19. As soon as the first results of a given step are known, researchers go straight to the next one. From the start of the research work to the beginning of clinical trials, which usually takes several years, it took only four months for SARS-CoV-2. Funding obviously facilitated this rapid development. In normal times, pharmaceutical companies start the production when they are sure they can put their product on the market. Here, governments have invested so heavily, pre-ordered millions of doses and paid in cash hundreds of millions of dollars to produce vaccines yet not proven to be effective.[51]" The situation is really amazing as there is yet no guarantee that the vaccine will deliver protective immunity and that it is safe. The Russian government was probably gambling—to some extent at least—and probably could not resist launching a big PR operation. However, no independent study was at the time available to confirm the results of the Gamaleya institute's researchers.

True, the speed of development of the first Covid-19 vaccines was really incredible. As an example, Table 4.2 gives a time line with the key steps along the way of the development of Moderna's mRNA vaccine.

At the end of 2020, the soon-to-arrive vaccines are presented as "the" solution and the key to a new normal life. Despite all the uncertainties and risks concerning vaccines at that time, and the fact that in some countries the

---

[51]Bafoil P (2020, August 16) Covid-19: "Il est possible que le vaccin soit disponible d'ici à la fin de l'année," Journal du Dimanche, https://www.lejdd.fr/Societe/Covid-19-il-est-possible-que-le-vaccin-soit-disponible-dici-a-la-fin-de-lannee-3985918.

opposition to vaccines is quite fierce, almost all the media across the world turned out to be crazy about them.

All in all, the Covid-19 story shows that all spheres of society have been under heavy pressure and that, most likely, no one will come out of this global crisis unaffected. Science will be no exception, which, despite its rigorous methods, has been seriously challenged by the coronavirus tsunami. The pandemic has forced the media to shine a spotlight on the work of the scientific community and has illuminated a dark side of science, highlighting the grassroots aspects of today's research. However, public opinion surveys carried out during the pandemic suggest that the image of science has not been tarnished—at least during 2020. In Germany and in the United States, polls have even shown an increase in confidence in science of 20% and 10% respectively during the first half of the year. France maintains a status quo, and surveys confirm the lessons of previous health crises: French people have a very stable and very positive attitude towards science. Thus, 77% of the surveyed population felt that they had neither more nor less confidence in science than before. A result that some will consider disappointing, given the strong mobilisation of researchers during the crisis, but which shows on the other hand that the image of science has not been penalised by the scientific issues which have been brought to the public's attention and have become media controversies.

In this context of infodemic, a question often comes up from the science world: how can we get science across? How can researchers communicate? How do we avoid succumbing to fake news and conspiracy theories? How do we communicate certainties in a climate of uncertainty? How do we respond to the discourse spread by the heavy machinery of the media given the worrying politicisation of scientific mediation that has so marked this pandemic? What can we do when our leaders are suspicious of science and scientists insist on the extent of our ignorance? The public no longer knows where to turn. Listening to the radio in the morning, I discover that the number of cases has exceeded "the threshold of 10 000." Then things are put into perspective by an epidemiologist who states that the authorities' discourse is "anxiety-provoking" and has shifted away from reality as the lethality of SARS-CoV-2 is now declining. Then my confusion continues as a tweet claims, with the help of relevant graphics, that the worst is yet to come. The next day, my favourite media outlet announces that the number of cases has "started to decline," with 25 182 cases against 25 315 the day before. Is a drop of 133 cases in one day really significant? That's half a percent… So why is the media making this point? And I am really at a loss when watching a TV debate bringing together doctors warning of the exponential increase of

ICU admissions and epidemiologists saying that the current wave is already declining. Transparency is good, but the proliferation of detailed data may also feed fear. For several months, we have been informed of successive mutations of SARS-CoV-2 via an almost *live feed*. This negative environment was the backdrop of the first vaccinations, in France in particular, where authorities advancing cautiously, made much of the multiple precautions they were taking, the side-effects of the vaccine and the fact that only a few people had yet been vaccinated, etc. A serious strategic error.

However, not everything is quite so bleak. A study conducted by British scientists shows that UK journalists have managed quite well with the huge influx of numbers and statistics that Covid-19 has brought into our daily lives and daily news like never before.[52] According to the study coordinator, An Nguyen, "British journalism – or at least a substantial part of it – has navigated the chaos rather well to identify patterns and to bring order out of the disorder." In any case, the scientists' findings provide vital evidence against the stereotyped image of journalism as a number-phobic, statistically incompetent profession. We have also seen the Italian and French media regain credibility during the pandemic.

Fighting against disinformation and suppressing misinformation is easier to say than to do. Baseless claims often spread faster than facts. Many medical professionals have found themselves combating the coronavirus on two fronts—in the hospital and on the Internet. Furthermore, purveyors of misinformation have no scruples about using insults and threats. There is no quick-fix solution. It's not easy to ensure that content is properly moderated. China's model of using selective censorship without transparency and accountability has turned the Chinese internet into a hotbed of both distorted information and ultra-nationalistic narratives. A China-style aggressive regulation of online content is no solution for the Covid-19 infodemic. The top priority must remain the quality of the information. This is the only solution against the propaganda, misinformation and disinformation, which is propagated by the media as well as politicians and widely disseminated on social networks. As highlighted by Herbert Lin and Harold Trinkunas: "Government leaders must respect factual information. It is clear that anti-intellectual populist leaders were inclined to downplay and minimise the risk of the pandemic for political ends have contributed to the current infodemic in

[52]Nguyen A et al. (2021, January 27) Reporting from a Statistical Chaos: Journalistic Lessons from the First Year of Covid-19 Data and Science in the News, https://www.bournemouth.ac.uk/news/2021-01-27/reporting-lessons-journalists-first-year-Covid-19-data-science-news.

ways that are both inimical to democracy and dangerous to public health. In democracies, at least, voters can change their leaders.[53]"

In my opinion, the academic and medical community should take the initiative. There is need and scope for an independent, professional and international news agency specialised in health, involving doctors, academics, researchers, care workers and journalists in a multidisciplinary perspective in order to combine scientific and medical expertise and professional communication according to a five-point strategy:

1. Bringing to the fore the information and expertise off the scientific and medical communities and communicating this information not via politicians but via sources that are highly and broadly trusted and demonstrably help people understand the crisis, as recommended by recent reports[54];
2. Providing comprehensive and high-quality information, scientifically and medically validated, and developing a method of communication that respects the public's capacity to understand the issues (relevant scientific references, data and technical notes, etc.). This would be the key to success: a high credibility means a virtual circle—news will attract new information upstream and media reports downstream;
3. Appointing a spokesperson and promoting the news produced by the network as well as its members (opinion polls confirm the positive image of scientists and doctors in public opinion, as opposed to politicians);
4. Using professional communication tools (which means being active on social networks, organising press briefings/points, developing mailing lists, producing visuals etc.);
5. Ensuring the scientific and ethical independence of the network and maintaining distance from private, official and governmental bodies. This means in particular securing funding from non-private and non-government sources (foundations, international organisations, donations, etc.).

Shortly after Joe Biden became President, officials called for a professional science communication initiative to be developed in order to address

---

[53]Lin H and Trinkunas H (2020, September 10) The Covid-19 infodemic: What can be done about the infectious spread of misinformation and disinformation, Bulletin of Atomic Scientists, https://the bulletin.org/2020/09/the-Covid-19-infodemic-what-can-be-done-about-the-infectious-spread-of-misinf ormation-and-disinformation/?utm_source=Newsletter&utm_medium=Email&utm_campaign=Thursd ayNewsletter09102020&utm_content=DisruptiveTechnology_Infodemic_09092020.
[54]Nielsen RK, Fletcher R, Kalogeropoulos A and Simon F (2020, 27 October 2020), Communications in the coronavirus crisis: lessons for the second wave, https://reutersinstitute.politics.ox.ac.uk/communications-coronavirus-crisis-lessons-second-wave.

the pandemic, relying on a "trusted public health source.[55]" In other countries, scientists, journalists and doctors have expressed the same concerns. But despite all possible efforts, every government communication will remain erratic as long as the decision process is short-circuited by the administration(s) in charge. We have seen many examples of last-minute decisions taken at the highest level in opposition to the scientific bodies, e.g., in France (opening of schools, third lockdown, etc.), in United Kingdom (herd immunity, late lockdowns , etc.) and in the United States (no federal lockdown, CDC's testing procedures, etc.). Public information will continue to be undermined until politicians, officials and scientists are able to forge a long-term, constructive and unambiguous relationship. This will undoubtedly help build mutual respect, putting more emphasis on the public's education and avoiding public health measures being guided by political positions. This is by the way what Joe Biden stressed on the day after his swearing-in, on January 21, 2021: "Our national strategy is comprehensive. It's based on science, not politics. It's based on truth, not denial, and it's detailed," he said. It is time now to go beyond good intentions.

Fatal and Political Mistakes

With vaccines rolling-out in an increasing number of countries, we may hope that the Covid-19 epidemic is under control. But it will take time to heal the wounds caused by this first global pandemic. Although far from the deadliest, this outbreak will be a landmark—at least until a threatening new virus arrives. For sure, this is not the last time we will hear about epidemics. There will be more to come and there will be worse to come.

History reminds us of those pandemics which were more fatal than Covid-19, such as the Black Death, which in the Middle Ages decimated half of the European population. More recently, there was notably the "Spanish flu" of 1918–1920 (which, despite its name, did not have anything to do with Spain), having caused nearly 40 million deaths worldwide (about two percent of the total population at the time) or the "Hong Kong flu," responsible in 1968 for one million deaths worldwide—much less than Covid-19.

But this crisis is different. Of course, it is first and foremost the consequence of a particularly dreadful viral epidemic, which has plunged the

---

[55]M. Field, Biden should focus on science communication as his administration seeks to tame the pandemic, Bulletin of the Atomic Scientists, January 19, 2021, https://thebulletin.org/premium/2021-01/biden-should-focus-on-science-communication-as-his-administration-seeks-to-tame-the-pandemic/?utm_source=Newsletter&utm_medium=Email&utm_campaign=ThursdayNewsletter01212021&utm_content=DisruptiveTechnologies_ScienceComm_01192021.

whole world into a long period of "staying at home" and an unprecedented economic recession—just like a war. The epidemic also revealed the government's unpreparedness in several countries. Mismanagement and miscommunication happened was rife. Undisclosed reports, data manipulation, lies, corruption: governments have used the complete set of tools to touch-up their image and make the situation look better than it is. Why not tell people the truth? The truth is that democratic leaders fear public opinion more than the coronavirus.

Covid-19 even highlighted problems which qualify as criminal practices,[56] such as disorganisation and politicisation of health systems, theft and concealment of medical equipment, misappropriation of public funds, etc. On each of the five continents, leaders have come under criticism and are accountable for the overall poor management of the crisis—or the crises should we say.

In their defence, we have to note that the pandemic raised many controversies, including in the scientific community. In many countries, supporters and opponents of restriction measures clashed, sometimes with violent protests, such as in Spain. In the United States, entrepreneur Elon Musk called lockdown a "fascist practice." In Europe, the emotion created by the epidemic will go down in history, according to Renaud Girard and Jean-Loup Bonnamy, as an example of "collective psychosis.[57]" The accumulation of conflictual issues (masks, hydroxychloroquine, lockdown, etc.) has undermined public confidence. In the United States, the CDC has been accused of lying for having discouraged the use of masks. During the shortage period, the U.S. health authorities decided to allocate masks as a priority to healthcare workers but the message was misinterpreted. Right up until the end of Donald Trump's mandate, communication from the U.S. administration has been abysmal. The first duty of government is to protect its citizens.

Under these conditions, the public no longer knows which way to turn… While European citizens remain confident in science, an opinion poll carried out in April 2020 simultaneously in France, the United Kingdom and Germany showed that the crisis had deteriorated public trust in their respective leaders and institutions, particularly in France. To describe the way in which their government is handling the crisis, 38% of the French interviewees chose the word "unpreparedness," while the Germans and the British preferred to speak, for respectively 44% and 38% of them, about "responsibility." The French believe at 84% that "the government will have to be

---

[56]These include: disorganised and politicised health systems e.g. in Lombardy, theft of protective masks in many countries, by a member of the staff of a Parisian hospital, assault on caregivers, etc.

[57]Girard R and Bonnamy JL (2020) Quand la psychose fait dérailler le monde (Tracts n° 21), Paris, Gallimard.

accountable," and 80% of them that there were "mistakes made by some members of the government." "The first consequence of the coronavirus crisis is the discrediting of the French State," writes the Paris-based correspondent of the Berlin daily *Die Tageszeitung* on May 2.[58] As Richard Horton sums up: "Covid-19 is a catastrophic failure of Western governments.[59]" An analysis which is confirmed, in a more polished way, by a WHO panel that highlights the responsibility of governments and public health organisations around the world for responding slowly and ineffectually to the outbreak. Faulty assumptions, ineffective planning, inefficient science alert system, major weaknesses in the global supply chain and sluggish responses helped fuel the pandemic, the experts conclude.[60]

However, as we have seen, a few governments are making a successful exit from the crisis. Some relied on technology to quickly isolate patients (tests, applications for contact tracing) while others encouraged individual responsibility or imposed social distancing without any delay. Germany's management of the epidemic confirms what the sociologist and anthropologist Paul Richards observed in Africa about the Ebola virus: it is, according to him, the awareness of the populations and the preventive modification of hygiene habits that prevented the outbreak from growing exponentially.[61]

The Covid-19 pandemic has been able, at the very least, to point to the huge fragility of our civilisation, dominated by the pervasive idea that scientific progress is driving back diseases all over the world. Science points out that in reality the opposite is true: new viruses and more resistant bacteria will continue to appear. Beyond that, the pandemic has emphasised the areas of our societies that are dysfunctional, which permanently operate on a "just-in-time" mode and tend to overlook the crises that undermine their common basis such as climate change, social decay, political discredit, overall debt, etc. Our way of life is still very far from being sustainable… As Dr Tedros wrote in an op-ed published in *Le Monde*: "We learned from the pandemic that health cannot be considered as a simple by-product of development nor as a cost that should be under control—it is an essential investment in more productive, resilient and stable economies and societies. […] A vaccine will

---

[58] Balmer R (2020, May 2) Die neue Maginot-Linie, TAZ, https://taz.de/Coronavirus-in-Frankreich/!5679530/.

[59] Morin H and Benkimoun P (2020, June 20) « Le Covid-19 montre une faillite catastrophique des gouvernements occidentaux», Le Monde, https://www.lemonde.fr/sciences/article/2020/06/20/richard-horton-le-Covid-19-montre-une-faillite-catastrophique-des-gouvernements-occidentaux_6043590_1650684.html.

[60] The Independent Panel for Pandemic Preparedness and Response (2021) Second report on progress, https://theindependentpanel.org/.

[61] Richards P (2016) *Ebola. How a People's Science Helped End an Epidemic*, University of Chicago Press.

help end the pandemic, but it will not address the weaknesses that cause it. There is no vaccine against poverty, hunger, climate change or inequality".[62]

It has been said that global capitalism is going to collapse like a house of cards when Covid-19 is under control. The only thing we can take for granted—and this is not new—is that the current liberal model of industrial and commercial development will be put into question, as production modes and consumption patterns will have to take into account global change and bring us towards a circular economy. However, a rapid and massive decrease in globalisation seems unlikely. American essayist Naomi Klein showed that crises are used by governments to consolidate their power and impose unpopular market changes to the detriment of the political and cultural values on which our societies are based.[63] This is not taking into account the considerable inertia of the system, as most companies will not give up the advantages procured by international production lines in terms of costs, competitiveness and profitability. Even during the pandemic, France rushed to order face masks from China! The sad point here is that globalisation, to many, is only seen as an opening up to competition, while it is also access to global cooperation. The crisis highlighted the weakness of democratic regimes and the power of technocapitalism, embodied by the global pharmaceutical industry.

In this unprecedented crisis, one thing struck me: the decline in international cooperation. Although the world suffered a global pandemic, countries turned in on themselves. Of course, we have seen incredible and self-less movements of generosity and solidarity everywhere: retired doctors and nurses returning to work; millions of people volunteering to help; restaurant chefs cooking meals for exhausted medical staff; luxury fashion brands manufacturing medical gowns; airlines companies offering their business jets; can manufacturers producing ventilators; etc. Behind the scenes, the scientific community was working hard to develop treatments for the disease and, as always, collaborating across borders. As during World War II, science has always been a link between countries in conflict. However, as much as World War II reinforced the military-industrial complex, Covid-19 boosted the politico-corporate complex, with companies across the world depending significantly on government funding and favourable authorisations. Among others, the U.S. financed the Warp Speed operation and the European Commission pre-ordered from pharmaceutical laboratories close to

---

[62]Adhanom Ghebreyesus T (2021, January 7) « Il n'y a pas de vaccin contre la pauvreté, la faim, le changement climatique ou l'inégalité», Le Monde, https://www.lemonde.fr/idees/article/2021/01/07/tedros-adhanom-ghebreyesus-directeur-de-l-oms-il-n-y-a-pas-de-vaccin-contre-la-pauvrete-la-faim-le-changement-climatique-ou-l-inegalite_6065465_3232.html.

[63]Klein N (2008) The Shock Strategy, London, Penguin Press.

two billion vaccine doses, if we take into account the options signed for additional doses. In total, this represents funding worth several tens of billions of euros worldwide.

But international exchanges remained limited, even if some countries gave examples of mutual aid and transnational solidarity. The only exception is the European Union. "Solidarity," recalled the President of the European Commission, Ursula von der Leyen, "is at the very heart of Europe and this is what will allow the EU to be born again.[64]" Europe set an example by deciding on April 3, 2020 to facilitate the transfer of infected patients to member states where ICU beds were available to alleviate the pressure on overcrowded hospitals. By relaxing its budgetary rules to allow member states to support their businesses and help people in need, the EU made it possible to release 2.8 trillion euros to fight the crisis—more than anywhere else in the world. Mme von der Leyen called for a "Marshall Plan" for the Union. The message circulated around the European capitals before and on May 27, 2020 the President presented to the 27 Heads of State an ambitious recovery plan financed by loans of 750 billion euros raised by the Commission—an initiative described as "historic." However, on the Old Continent, several leaders expressed doubts about the future of European integration and some even announced the end of the European *Union*. A big mistake of our governments has been to turn inward and to look only at the short term. This was even a fatal error which cost many lives. As the virus had become global, solutions needed to be global as well.

Even from the outset, WHO was very clear on this matter: "We must work together, united on all fronts," declared Dr Hans Kluge on March 31, 2020 in Copenhagen[65]: "This is a global crisis requiring a global, coordinated response. Leaders at the highest levels of government and industry need to work together to provide guidance, agree on strategy, devise protocols, and protect people, now as well as in the aftermath of the crisis." Turning inward would be a serious mistake. The difficulties we have been through confirm that we need not *less* international cooperation, but *more*.

I have been working on the international ITER project, which is building the world's largest experimental fusion reactor in the south of France. What is most remarkable is that this vast cooperation, unique in the world, makes it

---

[64]von der Leyen U (2020, April 4) Cette solidarité est au cœur même de l'Europe et c'est ce qui va lui permettre de renaître, Le Monde, https://www.lemonde.fr/idees/article/2020/04/04/ursula-von-der-leyen-cette-solidarite-est-au-c-ur-meme-de-l-europe-et-c-est-ce-qui-va-lui-permettre-de-renaitre_6035513_3232.html.

[65]WHO Statement—Countries must work together as Covid-19 pandemic accelerates, https://www.euro.who.int/en/media-centre/sections/statements/2020/statement-countries-must-work-together-as-Covid-19-pandemic-accelerates.

possible to achieve together what none of the 35 participating countries could do alone. Only international collaboration can tackle global challenges.

I think one of the best initiatives politicians could take is to provide a global governance for quality of life, which encompasses health, environment and access to minimum resources. This would ensure the well-being of all generations, present and future, as well as ensuring that the preservation of essential resources necessary for humanity remains a central focus for political action. The many requests for a universal access to Covid-19 vaccines just express that quality of life should be recognised as a global public good and a fundamental human right. We share illnesses, but not always treatments. An international approach would be one more step towards the recognition of this universal heritage to which we all belong. We are already heading down this path. There is a lot of international cooperation in the fields of science, health and the environment. This contributed to the improvement of the global life expectancy, which has increased from 66 to 71 years (average females and males) since 2000. But there are still significant and unacceptable disparities in a globalised world. So, a first step would be the organisation of world conferences—much like the United Nations Climate Change COP conferences—to allow world leaders to take the pulse of mankind and undergo a health check together in order to identify priority issues. These international events are not a goal in themselves but a means to coordinate and take action together. I have said and written many times that these big gatherings are far from the final solution. We can see this clearly during these events, where there seems to be a big difference between the messages in the speeches and the real intentions of the so-called decision-makers. The reality of politics or just hypocrisy? Superficial decision or genuine indecision? Are policy-makers addressing the real problems in an efficient and comprehensive way? It is for you to judge. The fact remains that these world events contribute to progress, too slowly of course but every little count in raising awareness, sharing knowledge and stimulating transnational cooperation. These events also provide an operational base for new missions.

In theory, everything is already in place. WHO provides leading international expertise in the field of health. Its official objective is indeed to help "all peoples attain the highest possible level of health." In 1995, the organisation established a division on Emerging and other Communicable Diseases Surveillance and Control and in 2000 set up the Global Outbreak Alert and Response Network (GOARN) to help countries manage epidemiological and pandemic crises. A leader in global health, WHO advises nearly 200 countries on disease and epidemic management. However, it is regularly exposed to criticism. While countries accept the organisation's expertise and advice, many want to be in control. And the point is, the organisation has no coercive power on the member states that refuse to cooperate with it. Let's be clear: some states want WHO to remain weak because health is an eminently political subject and remains a national competence. The life of international organisations is difficult! Most of them have been created following wars or major crises (such as the League of Nations in 1919, the United Nations in 1945, the European Commission in 1957, etc.), then they have to struggle to receive adequate resources from political authorities. The European Union also has the means to improve public health and public health security. Despite President Macron's remarks calling for the creation of a "Europe of health," article 168 of the Treaty on the Functioning of the European Union (TFEU) entitled "Public health" gives European leaders a framework of action. Health remains essentially the responsibility of Member States, but the Union complements national policies and encourages cooperation between countries in the field of public health. The foundations of a Europe of health are already there. Furthermore, the 27 EU member states have called for strengthening the powers and resources of WHO.

When Bill Gates said in 2015 that "We are not prepared for the next outbreak" and suggested creating an army of specialists from many disciplines to meet whatever crisis or epidemic might arise, millions of people viewed his "TED talk" but nobody in power heard the message. Only a salutary shock could raise a global awareness of the need—and the urgency—to work together. The coronavirus pandemic has clearly shown that, going beyond borders, global management is the common-sense solution. "In this pandemic," Bill Gates says on April 12, 2020, "We are all connected. Our response must be, too.[66]" It is no coincidence that Covid-19 spread at a time when the world is ruled by Donald Trump , Jair Bolsonaro, Narendra Modi and Vladimir Putin and when nationalists are gaining ground, pushed in particular by Brexit in the United Kingdom and the Five Star Movement in Italy. They are all holding up the mirage of a quick and easy solution—close the borders and solve the problem themselves. This was a fatal mistake. The threats of global warming and of present and future epidemics will be contained and eventually sorted out only if we act together and foster better collaboration between all countries and with all organisations, including the active participation of the general public. We have seen the limits of national strategies when France lifted its lockdown in the summer of 2020. From mid-June, Guyana, a French overseas territory, was hardly hit by the epidemic and

---

[66]Gates B (2020, April 12) Coronavirus: World needs a global approach to fight, The Australian, https://www.theaustralian.com.au/world/coronavirus-world-needs-a-global-approach-to-fight-bill-gates/news-story/49ee4ad03e03177364e3a85c19216e28.

decided to go into lockdown to protect its population from the spreading epidemic in the Latin America neighbour countries, and Brazil in particular. Our national borders are almost transparent when it comes to a global pandemic.

There is much more to be gained by working together, and that is also the core mission of WHO. This could lead in particular to the harmonisation of testing methods and counting cases, both those infected and recovered, and deaths. We're not even there! Boosting vaccine research is another activity that requires coordination, pooling funding, and sharing results across the globe.

We should go even further. As I put it, health must be recognised as a global public good. Health and social assistance systems cannot be developed by following market forces and a genuine equal opportunities policy should guarantee health coverage for all. Economist Thomas Piketty proposes every inhabitant of the planet is allocated a "minimum capital endowment" which would include equal access to education and health.[67] This would be financed by establishing a fair and progressive tax system, which could lead to redistributing part of the tax revenues paid by the most prosperous economic players. The United Nations effectively protects the world heritage—but should we not consider the whole of mankind among the wonders of the world?

The WHO should not be given less responsibility, but more power. The problem is the same as far as environment and global warming are concerned: what happens in a country has negative consequences far beyond its own borders. Wherever it is, the coronavirus threatens us all. The French President has the right to impose lockdown on his entire country. Why can't the Director General of WHO do the same for the sake of all humanity?

When will national politicians recognise that they fall short? This is neither a personal criticism nor because I suspect them all of incompetence. My question just follows up on the fact that today's major problems can only be solved if they are solved for the entire planet and not for just one country in particular. I am not advocating a technocratic, supranational and elitist model. Let us just apply at global scale the practices used today to reinforce democracies: involvement of citizens in decisions, decentralisation of political power, integration of the scientific developments. Health, environment, climate, energy: all these issues must be tackled together and not by each country in his own corner. To quote Kofi Annan, former United Nations'

---

[67] Piketty T (2020, April 10) L'urgence absolue est de prendre la mesure de la crise en cours et de tout faire pour éviter le pire, Le Monde, https://www.lemonde.fr/idees/article/2020/04/10/thomas-piketty-l-urgence-absolue-est-de-prendre-la-mesure-de-la-crise-en-cours-et-de-tout-faire-pour-eviter-le-pire_6036282_3232.html.

Secretary-General in 2001: "The only route that offers any hope of a better future for all humanity is that of cooperation and partnership." We have to move together towards effective global governance which, by transgressing national borders, will lead to improving global quality of life. This change is already underway in Europe, where the decisions are taken more and more at transnational and regional levels, the national level eventually becoming an empty shell. There will be no obvious improvement in the current situation if individual states do not cede some prerogatives of their national sovereignty in favour of better access to health—the number one global public good. I do not have a turnkey solution to hand but there are clearly several avenues to explore, if the political will is still there after the crisis. Our village is now a planet.

Scientific and Political Education

For its size, SARS-CoV-2 has made an enormous impact. A virus with just ten genes has turned the world upside down by triggering a global health, social, political, economic and humanitarian crisis. A scientific crisis as well. The fight against the virus and the disease has brought to light the way research works, the way science is done, the way scientists interact and compete with each other and with society. This has highlighted scientific disputes, quarrels between experts and institutional dysfunction. Science is life! Above all, the coronavirus outbreak has shown that science under pressure does not perform at its best: downplaying the outbreak, wrong vaccine predictions, peer-reviewed publications retracted after a few days, short-circuiting of clinical trials, research supported for political reasons, scientific vagaries etc.—we have seen many excesses and errors brought about by this exceptional situation. Covid-19 was the first global pandemic managed *live*, in real time, with instant feedback and information from the four corners of the world. The public also discovered "science by press release" as many research organisations and businesses were also actively carrying out self-promotion and marketing. With the help of social networks and the global media, we learn everything about the virus and the status of public health, on a day-by-day basis. And above all, we learn that we still have a lot to learn.

During a crisis, extremes move closer together, for better or for worse. We have seen initiatives of true generosity alongside acts of pure selfishness; politicians taking their responsibilities seriously whilst scientists chase opportunities. Although I am not agreeing with the relativism around us, what we have seen since the start of the pandemic is that science is a human activity like any other, where conquest of power sometimes takes precedence over the quest for knowledge. It is also a real *market*, offering careers and innovations to the highest *bidders*. Nothing new, one may say: Covid-19 confirmed that science is not the absolute bulwark or the all-powerful weapon that may be used to rationalise politics and politicise science. Our countries, which rely heavily on science for the future-proofing, also have complex relations with it. It is interesting to see how the two countries which are today the biggest investors in scientific research dealt with the coronavirus outbreak: China, which scrupulously followed its epidemiologists and infectious disease specialists, implemented a policing strategy to apply their recommendations and counter the epidemic while the United States, which handled the crisis in an inconsistent and, say, irresponsible way, went against the scientific community. Do we have a kind of schizophrenic relationship with science?

The Covid-19 pandemic has highlighted our difficult relationship with numbers, math and, more generally, science. When the value of the now famous R_t was a bit higher than one, authorities often called for "vigilance" or decided "to monitor the situation" underestimating that an exponential—and therefore explosive—development was underway. Our leaders had the best expertise and the alerts worked well. But the lights did not turn red. The numbers don't speak and, for most of us, they don't actually communicate anything. Numbers generate many misunderstandings, as French mathematician Stella Baruk has shown, in particular because the general level of competence in mathematics is decreasing, leading to widespread "innumeracy," the mathematical counterpart of illiteracy. Beyond that, we still have a contrasting relationship with scientific knowledge and those who embody it. We all have a reluctance to admit our ignorance. For cultural reasons in part, we are reluctant to recognise our incompetence, even though this can also trigger, under certain conditions, a positive or even creative process.[68] Science and technology education is a priority now more than ever. Just as society obliges new drivers to pass an examination to get a driving license, a "science test" could help us navigate a technological world and better understand what is scientifically possible and what is not. The problem is also societal: this pandemic raises (again) the question of the interaction between science and society and the conditions for establishing a long-term dialogue and not just a one-off cooperation, whenever a crisis emerges and catches everyone off guard. Science reassures us because scientists still embody the myth of the truth. On January 12, 2021, on the French *TF1* news, journalist Gilles Bouleau was asking Jean-François Delfraissy for more information on the new

---

[68]Claessens M (2013) Petit éloge de l'incompétence, Paris, Quae.

SARS-CoV-2 variant: "Tell us, you are the scientist." However, who needs this logorrhoea of statistics which follow one after another all day long? Who needs daily information on vaccines in development? Do we better understand the disease by listening to these interviews of doctors and researchers that the media rehash all the time? Shouldn't we become a little fatalistic, take it upon ourselves and show solidarity with Nature by deciding, in bad weather, to stay home and not to go on vacation?

Unless our society has a healthier and a more honest approach to the world, such crises will be inevitable. It is about our values, and therefore our lifestyle. It is about society, and therefore about the respect we have for others. It is also a question of science, and therefore of what we want to do with knowledge. Let's say it once and for all: science is not the magic trick that can solve the planet's problems. This is not its role, although some discoveries and innovations may create wonders. It is up to scientists not to leave any ambiguity. From the first days of the pandemic, many voices heralded the vaccine as the final solution to Covid-19. Scientific research cannot remove our individual responsibility. Growing urbanisation, globalised trade and mass tourism are also contributing to the emergence and spread of new viruses. Let's stop calling on technoscience to save the day as soon as a problem arises! Our lifestyle and ways of thinking are also, at least in part, at the origin of this global crisis.

We are still paying the price for the success of the Manhattan project which, during World War II, allowed the United States to win the atomic bomb race and the allies to liberate the world from the Nazis' domination. This success also marked the start of the marriage of science and politics. On August 6, 1945, as the first A-bomb exploded in Hiroshima, *Le Monde* published a front-page article of which the headline read: "A scientific revolution." As a result, the project has deeply permeated the functioning of our societies. It profoundly influenced scientific policy in developed countries and led to the paradigm of scientific research being the engine that drives the development of our economies and societies. In particular, Manhattan inspired President Franklin Roosevelt's scientific advisor, Vannevar Bush. He designed a linear model that assumed a direct link between scientific knowledge and socio-economic development, through the successive stages of research, invention and innovation. This model, which is also based on the idea that fundamental research must be stimulated through the availability of resources, still influences the scientific policy of industrialised countries.

Since then, science has, for many of us, been reduced to just the sword arm of our society: curing diseases, creating innovation, ensuring a more comfortable life. Since the second half of the twentieth century, science has been hijacked in a sense: the quest for knowledge and the study of Nature became secondary objectives. This is reflected in particular in the growing share of research funded by industry. Modern science has also become political, as a result of trade-offs over the resources devoted to it and its institutional implementation.

This hard trend coupled with funding shortages requires researchers to outbid their contribution to society's problems and sometimes to promise miracles. A cure for cancer? A few million euros and we'll give you a solution! A technique for energy storage? Hire researchers and give them the necessary resources! A vaccine against colds? Go and support the pharmaceutical industry! However, on these three issues, we are still awaiting the promised solutions…

Covid-19 is now another good example. The response to the pandemic, in 2020 at least, was not a scientific issue. In the absence of any treatment and vaccine, the priority was to achieve physical distancing. While technology helped identify those infected, the effectiveness of isolation—lockdown or quarantine—was wisdom from the Middle Ages. The production of vaccines is another technology which the pharmaceutical industry is improving continuously—as we have seen. The development of mRNA vaccines is an undeniable success but the technique had been in development for over twenty years. It was the funding and resources allocated to it that made the difference. On April 2, 2020, Anthony Fauci estimated that a vaccine could not arrive for 12 to 18 months. Experts said a quick approval could be risky. But the industry took up the challenge: the International Federation of Pharmaceutical Manufacturers and Associations (IFPMA) claimed on March 19, 2020 "an unprecedented level of collaboration taking place across the industry."

Techno-capitalism is the big winner of the pandemic. The medical industry and governments saw their powers reinforced. Managing the outbreak in early 2020 was not up to scientists. Managing a lockdown is more a matter of police decisions than a scientific process. Science's most valuable contribution is to develop and bring in robust knowledge. But this takes many years. Science does not do quick fixes and repairs on demand. During the crisis, epidemiologists underlined the difficulty of the exercise consisting of moving from conventional science, which aims to develop our scientific knowledge, to "in real time" epidemiology, which aims to advise decision-makers, a move that forced researchers to radically change the way they communicated—from peer-reviewed publications to mainstream media. In early 2020, scientists provided their governments with warnings, information and simulations. Then it was up to the decision-takers to decide! Researchers and doctors should have been clear about this. This ambiguity has generated many misunderstandings and false hopes. In a way, scientists were incompetent. It was not their job. From this perspective, the much-awaited production of vaccines against Covid-19, which is of course very good news, might help scientists only in the short-term as it reinforces the idea that scientific research is like a magic trick that can solve any problem. We should stress on the fact that the technology used to produce messenger RNA vaccines is the culmination of more than twenty years of fundamental research.

Instead, scientific councils and advisory boards have had to manage police decisions and determine the conditions for achieving distancing and isolation. The science advice hastily put in place by governments to help them manage the pandemic was, at best, as much use as sticking a plaster on a wooden leg or, at worst, a smokescreen for the public. In many countries, we saw eminent scientists talking about the usefulness of curfews and partial lockdowns. Have we forgotten how exactly science works? I couldn't help but smile when I heard this renowned epidemiologist answering a journalist's question: "Is it a good idea to shut florists during lockdown?" Or when this famous virologist vigorously opposed the opening of hair dressers. Scientists should not have been drawn into the pseudo-scientific management of the crisis promoted by their government.

At the start of the pandemic, in January 2020, the first priority was the development of testing techniques, diagnostic methods and tracing applications. Scientists and engineers did a good job. However, we missed the international harmonisation which would have made it possible to measure the progression or decline of the outbreak across the world, validly compare regions and countries, and identify good practices. This work is not yet complete. Ensuring that Covid-19 tests work 100% every single time and in the same way for every person really matters to people's lives. And all the countries of the world will benefit from this, especially the developing countries. Making science more robust, reliable and transparent increases the performance of the research system as well as society's trust in science.

Scientific research is slow but progress builds over the long-term. Research will provide the answers to the fundamental questions that will determine our ability to react to future pandemics: why do certain viruses cross interspecies barriers? What are the most pathogenic protein structures? How does our immune system interact with coronaviruses? But the contribution of science is happening in the long term and is closely related the advancement of knowledge. The testimonies presented in the previous pages have shown it clearly: scientists, who rightly give priority to the "evidence", are not in the best position to guide political decision-making. They are, moreover, no better placed, one may even say, to guide science. First, because inspiration cannot be controlled. And also because, nowadays, research is essentially progressing in the fields where, following a convergence of funding, epistemological reasons and political support, publications follow…

Covid-19 also confirmed that, during a crisis, governments put excessive pressure on science and scientists. There is a huge misunderstanding here which derives from the "Manhattan success." Politicians hope to see immediate results and expect the research system to be tuned to serve the crisis. The previous chapters showed many examples of governments manipulating administrative and scientific procedures to hastily approve unproved drugs and influence the statistics. Some research projects have been supported and funded for political reasons. There was also a lot of pressure on scientific publishers, funding agencies and regulatory bodies to accelerate the development and the approval of new products. The problem is that scientific research does not work like this. "When good science is suppressed by the medical-political complex, people die," warns Kamran Abbasi, the executive editor of the BMJ.[69]

We therefore have to reflect on what are the best conditions to promote this interaction between science and society in a sustainable manner. This requires the legitimacy of both political power and scientific expertise, as well as scientific power and political expertise. Ensuring this should allow constructive and long-term exchanges to be established, as well as full respect between political decision-making and scientific knowledge, which remain quite compartmentalised. I do not mean here that the advisory committees set up by governments are simply lip service.[70] But the information mechanisms and the structures put in place in all countries to inform elected politicians of science's most up-to-date information and the possible options are not enough—or no longer enough. True, many initiatives have been taken in recent years to strengthen the governance of technoscience. However, one may fear that the variety of these attempts and the diversity of the actions implemented may contribute to blurring the messages and diluting the resources.

The truth is, we have yet to find the right mechanism to make the most of science and guide political action. It would be wrong to stress the politicisation of science without acknowledging that science is already politicised. We were all struck by the indulgence of the scientific committees with regard to

---

[69]Abbasi K (2020, November 13) Covid-19: politicisation, "corruption," and suppression of science, BMJ 371, https://doi.org/10.1136/bmj.m4425.

[70]Bréchet Y (2018) Science and politics, Commentaire, nr 161.

the mistakes made by governments during the pandemic and the silence of the medical community about the risks and uncertainties of treatments and vaccines against Covid-19. Of course, politicians are exploiting science, but scientists are also exploiting politics to promote their own work and engage in a kind of "consumer marketing". In fact, science has always been inseparable from politics.[71]

Neither politicians nor scientists seem to accept this reality of our techno-scientific democracies. Our leaders, with rare exceptions, do not seem to consider it essential to ground the decisions they have to make on scientific evidence. In exceptional cases such as the current pandemic, cutting-edge scientific expertise is taken into account by the government, but mainly because it would be unacceptable were it not. If our rulers tend to stand somehow above the law, they cannot stand above science. On the other hand, many researchers are keen to emphasise, even with a certain pride, their distance from politics. Scientists still hope to regain control of the decisions of society, under the pretext that they act as the voice of reason. Quite rightly, politicians suspect that scientists are aiming to make decisions on their behalf. And scientists have a limitless disdain for politicians. Both spheres are inhabited by an instinctive distrust of the other. The conditions for genuine dialogue and respect are therefore not met. A situation for which the two camps bear some responsibility. And this difficulty is reinforced, in the event of a crisis, by the different time scales: leaders are obsessed with day-to-day action while scientists need to look at the long term.[72]

Far be it from me to argue that researchers are not interested in the problems of society. Indeed, the opposite prevails—in part to ensure their funding and also for trivial reasons. They enthusiastically respond to calls for projects to cure cancer, find clean energies and develop artificial intelligence. However, the main reason is sometimes that they are lacking funding to continue to research the subjects that fascinate them—which may have a rather distant link with the issues raised by society. There are other considerations, more specific to the very nature of research, which limit the contribution of scientists. Like the fact that, in most cases, researchers do not know exactly what they are looking for, and more importantly, what they are going to find. In addition, in their laboratory, researchers work on simplified models, which are sometimes only of academic interest. Real-world cases are appallingly

[71] Sabbagh U (2017, April 25) Science Has Always Been Inseparable from Politics, Scientific American, https://blogs.scientificamerican.com/guest-blog/science-has-always-been-inseparable-from-politics/.

[72] Roucaute D (2020, September 28) Epidémie de Covid-19: « Tous les gouvernements ont commis des erreurs», Le Monde, https://www.lemonde.fr/planete/article/2020/09/28/pandemie-de-Covid-19-tous-les-gouvernements-ont-commis-des-erreurs_6053898_3244.html.

complex! This situation creates huge expectations. Political pressures, media controversies and scientific vagaries are the other side of the coin. Above all, there is a qualitative difference between science and expertise. Being the world's top specialist of the carbon dioxide molecule won't automatically make you an expert in global warming. As Philippe Roqueplo has clearly shown, acting as an expert leads you inevitably to transgress the limits of scientific knowledge: "Think of nuclear energy, mad cow disease, genetic manipulation: it is deceptive to believe that one can be neutral on such questions! [...]. Expertise requires scientists to express convictions that go beyond their knowledge.[73]" Hence the need to verify the independence of experts and the absence of conflicting interests in particular with lobbies and private funding.

There is one initiative that has successfully contributed to build *sustainable* links between science and policy, and it is the IPCC, the Intergovernmental Panel on Climate Change, which publishes periodic reports and authoritative assessments on the whole issue of global warming. Created in 1988 by the World Meteorological Organization (WMO) and the United Nations Environmental Program, the panel, made up of representatives of the 194 participating governments, regularly takes stock of the physical scientific aspects of climate change, its global and regional impacts, as well as the options for adapting to and minimising them. Concretely, under its remit, the IPCC sets the scope of its expert working groups, elects the board that oversees them, and approves the summary for policymakers (SPM) that accompanies the extensive assessment reports, prepared by some thousands of volunteer scientists. That work earned the IPCC a lot of respect, as confirmed in particular by the Nobel Peace Prize awarded to the expert body in 2007.

Even if criticisms have been voiced about the IPCC and notably its impartiality, its mode of operation and its too slow and not very transparent way of communicating (in particular on the reports' errors and uncertainties), the intergovernmental network embodies a new global technoscientific governance by making scientists, economists and politicians work together. There is unanimity on this point. Supporters and opponents to climate change join together in recognising that this UN body has created a formidable striking force in support of the theses of anthropogenic warming. And we really need it! If, as the saying goes, success is measured by the number of enemies, the IPCC is unquestionably a great success today. Despite the many attempts made in recent years to discredit the value of its work, the international group has established itself as a model for technoscientific governance, in particular

---

[73]Roqueplo P (1997) Entre Savoir et décision, l'expertise scientifique, Paris, INRA.

by combining scientific and political skills in a policy- and decision-making perspective. It is therefore not surprising that the political impact of the IPCC goes beyond the climate framework and influences areas where science plays a major role.

Despite the IPCC's imperfections and very political profile, this peculiar structure plays an essential role in articulating the scientific and political worlds, which are then working together, interacting with each other and progressing jointly, even if there is no global consensus. French scientist Amy Dahan considers this structure as a suitable model for other international expertise projects.[74] For example, the "Intergovernmental Scientific and Policy Platform on Biodiversity and Ecosystem Services (IPBES)," established in 2012, is based on the same model. Created by 94 governments, this initiative provides assessment reports and aims to interface scientists and policy-makers on the state of biodiversity in the world and on ecosystem services. Our governments could draw inspiration from these experiences and set up an "IGEP", an international group of experts on pandemics to achieve a "co-construction" of expertise and decision-making in order to provide scientific grounds for political decisions and present research results in a suitable way for political action. These two spheres, as we saw during the pandemic, are still working in parallel words, which are joining forces, hastily, during one-off crises. In reality, the linear model research → expertise → decision is far from the most efficient. Historians and sociologists have shown that it is totally inadequate to account for the complex back and forth between fundamental knowledge and their applications. The IGEP could therefore put together the conditions for a common project, which the political world and the scientific community will make their own. Some may see this project as too much of a technocratic approach. They are not wrong, and I would recommend to include a "citizen science" component with, for example, participation of representatives from civil society. These convergences would be the strong point of the IGEP: a central place given to science in the definition and the resolution of global health risks by allowing the development of evidence-based policies and the involvement of society at large.

This group of international experts still has to be fleshed out, but it will be essential to guarantee its capacity for action and influence, both scientific and political, in order to support or complement WHO's mandate, which is to support and coordinate international health. No doubt it will have to integrate a permanent structure, because other epidemics and other health

---

[74]Dahan A (2007) Le changement climatique, une expertise entre science et politique, La revue pour l'histoire du CNRS, 16, 2007, https://doi.org/10.4000/histoire-cnrs.1543.

problems will follow. Above all, the IGEP would constitute a permanent support group and would perpetuate expertise in epidemiology and the fight against infectious diseases. First, it would take stock of this pandemic, in terms of scientific knowledge and political action. We have seen too often, with Covid-19, that groups formed during previous epidemics had been dissolved and that everything had to be restarted from scratch... In an ideal world, the IGEP could complement or even supplant the many national scientific committees which have been created for Covid-19. The multiplication of these committees and the limitation of their field of action at national levels are undoubtedly an obstacle to a rapid response, complicate coordination, deteriorate communication and prevents the implementation of coherent international strategies. What is at stake is the need to build on the best management practices of the Covid-19 crisis. But let's not be naïve: too often, in the real world, our leaders prefer to reinvent the wheel and stay in control. The price to pay is managing personal influences and rivalries as well as political pressures, often exacerbated when we work in small committees...

Interestingly, while I am finishing this book, on March 30, 2021, 25 world leaders, including Prime Minister Johnson, President Macron, Chancellor Merkel, European Council President Michel, and WHO Director-General Tedros Adhanom Ghebreyesus set about drafting a new international "pandemic treaty." Describing Covid-19 as "the biggest challenge to the global community since the 1940s" and "a stark and painful reminder that nobody is safe until everyone is safe," they argue that the world needs a global settlement similar to that formed after the Second World War to protect states in the wake of the pandemic.[75] The aim of the treaty is to ensure that the world learns the lessons of the Covid-19 pandemic and urge greater investments in preparedness by strengthening national, regional and global capacities: "No single government or multilateral agency can address this threat alone," the leaders state.

These initiatives are timely and fit nicely in setting up a suitable international framework which would be the "natural" basis to tackle planetary crises by providing a long-term interaction between politicians and scientists, a co-construction of the political action by scientific validation and a basis for international cooperation. This would create the conditions for genuine cooperation and mutual respect. Science cannot act piecemeal and provide quick fix solutions. Under these conditions, scientists would be allowed to

---

[75] Fischer L and Nuki P (2021, March 30) Exclusive: World leaders call for pandemic treaty, The Telegraph, https://www.telegraph.co.uk/politics/2021/03/29/exclusive-world-leaders-call-pandemic-treaty/.

focus on basic research and their contribution to the advancement of knowledge. This is what they do best. Our ignorance is limitless and calls for unreserved modesty. And it is ignorance—not knowledge—that is the true engine of science.

Bibliography

1. Bourdelais P (2003) Les Epidémies terrassées: Une histoire des pays riches. La Martinière, Paris
2. Caumes E (2020) Urgence sanitaire. Robert Laffont, Paris
3. Claessens M (2011) Allo la science Herman,Paris
4. Dahan A, Aykut S (2015) Gouverner le Climat ? Presses de Sciences Po, Paris, Vingt années de négociations internationales
5. Debré P, Gonzalez JP (2013) Vie et mort des épidémies. Odile Jacob, Paris
6. Dessler AE, Parson EA (2019) The science and politics of global climate change: a guide to the debate. Cambridge University Press, Cambridge
7. Fang F (2020) Wuhan diary. HarperVia, New York
8. Feyerabend P (1975) Against method: outline of an anarchistic theory of knowledge. Verso Books, New York
9. Firestein S (2012) Ignorance: how it drives science. Oxford University Press, Oxford
10. Girard R, Bonnamy JL (2020) Quand la psychose fait dérailler le monde (Tracts n° 21). Gallimard, Paris
11. Hirsch M (2020) L'énigme du nénuphar face au virus. Stock, Paris
12. Horton R (2020) The COVID-19 catastrophe. Polity Press, Cambridge
13. Kierzek G (2020) Coronavirus—comment se protéger ? Archipoche, Paris
14. Lamotte P, Libaert T, van Ypersele JP (2015) Une vie au cœur des turbulences climatiques. De Boeck
15. Lévy BH (2020) Ce Virus qui rend fou. Grasset, Paris
16. MacKenzie D (2020) COVID-19: the pandemic that never should have happened and how to stop the next one. Hachette Books, New York
17. Michel JD (2020) COVID : anatomie d'une crise sanitaire, humenSciences, Paris

18. Moore P (2015) Le Petit livre des grandes épidémies. Paris, Belin Sciences
19. Oldstone MBA (2020) Viruses, plagues, and history: past, present, and future. Oxford University Press, Oxford
20. Pestre D (2013) A contre-science. Politiques et savoirs des sociétés contemporaines, Paris, Le Seuil
21. Pestre D (2014) Le gouvernement des technosciences, Paris, La Découverte
22. Pialoux G (2020) Nous n'étions pas prêts—Carnet de bord par temps de coronavirus. JC Lattès, Paris
23. Raoult D (2020) Epidémies: vrais dangers et fausses alertes. Michel Lafon, Paris
24. Richards P (2016) Ebola. Chicago, University of Chicago Press, How a People's Science Helped End an Epidemic
25. Roqueplo P (1997) Entre savoir et décision, l'expertise scientifique. Inra Editions, Paris
26. Saxena SK (ed) (2020) Coronavirus disease 2019 (COVID-19). Springer, Singapore
27. Siegel M (2020) COVID: the politics of fear and the power of science. Turner Publishing Company, Nashville
28. Spinney L (2018) La Grande Tueuse—Comment la grippe espagnole a changé le monde, Paris, Albin Michel
29. Woodward B (2020) Rage, New York, Simon & Schuster

Index

Printed in the United States
by Baker & Taylor Publisher Services